D0103022

PHILIP CONKLING

DAVID D. PLATT

Editors

HOLDING GROUND

THE BEST OF *ISLAND JOURNAL*

1984 – 2004

PETER RALSTON

Island Journal Art Director

PAIGE GARLAND PARKER

Design

Editors: Philip Conkling and David Platt
Island Journal Art Director: Peter Ralston
Design & Production: Paige Garland Parker
Production Assistance: Charles G. Oldham
Map Graphics: Chris Brehme, Liv Dietrik
Printed in China by Four Colour Imports, Ltd.

Front Cover Photo: Peter Ralston
Back Cover Photo: Peter Ralston

Copyright: Island Institute 2004
Box 648, 386 Main Street
Rockland, ME 04841
www.islandinstitute.org

ISBN#: 0-942719-34-4

The Fish Wharf, Matinicus Island, George Bellows, 1916

MATINICUS
68°55′W — 43°52′N

Further out than a mainland eye
can see, it lies hull down in the mind,
an island that you can reach for, hazed
in the lightest airs, somewhere offshore
until it lifts across, and off,
the flat blue curve far out: a landfall
wavering more in time than distance.
No compass, no degrees and minutes,
can chart you back, who stand watch first
for No Man's Land, and a reef awash
where the swell breaks white like a whale.
The floodtide sets you, past the bell;
you remember now, and anchor close in,
in the hooked cove where fishhouses shine
in the new-paint sun. The shore is slant
granite slabs, blank windows looking out
from white clapboard, where the catwalks sag

seaward from door to front door; the plank
weathered, like bait tubs, or the thin pier
stilted back to black spruce. The churchspire,
high ashore, might be the landmark
for your bearing here. Or Matinicus Rock,
where (homing in on the diaphone
or light) you might come to find puffin
after a two-reef windward beat.
But anchored now, with a lobsterboat
astern, and the whole sea behind you,
you recognize your holding ground
and know what deviation swung
your compass back; set out along
the catwalk windows' opened frames,
Matinicus geraniums,
as for the twenty summers you forgot,
blaze like light buoys, each in a tin-can pot.

Philip Booth, *1986*

IT WAS *ISLAND JOURNAL* THAT GOT TO ME nearly twenty years ago. I had spent as much of my life as I could doing things here and there along the Maine coast, like exploring it in small boats, fighting off an oil refinery in Eastport, sinking summer roots on Vinalhaven, working in Portland. So the first *Island Journal* I ever saw really got my attention. It encapsulated the Maine coast and its island communities better and more completely than anything that has ever been produced by anyone, before or since. Period. And it has done that now for twenty years. What it captures once a year is not just the beauty, which is staggering, but also the history and the culture of the islands and the island communities, the gritty reality of "islandness" and the character of the people who live on them.

Philip Conkling once told me that it was Betsy Wyeth, wife of Andrew, who urged him and Peter Ralston to do the *Island Journal* in this fashion, to make it the best, of the highest quality possible, even if the fledgling Island Institute could scarcely afford it. Her advice was sound. *Holding Ground* is a distillation of twenty years of following that advice. David Platt, the Island Institute's editor and director of its various publications, most particularly *Working Waterfront*, has over the years drawn enormous talent to *Island Journal* in the arts, history, politics, natural science, current affairs and all the things having to do with islands and islandness. If the Island Institute did nothing else but publish *Island Journal* once a year, it would hold a place of honor among the greatest of Maine's nonprofit organizations.

It is therefore remarkable how much more the Island Institute does. It truly nourishes and helps sustain the fifteen island communities that still exist along the Maine coast, nurturing their schools, equipping them with island fellows (a very practical kind of island Peace Corps), marshalling political support for them, promoting communications among them, helping them plan their future and, frankly, celebrating them.

Philip Conkling and Peter Ralston, founders of the Institute, are far too modest and professional to blow their own horns, so perhaps, however inadequately, I can do it for them. The reservoir of talent that resides in these two men is simply remarkable. Peter's photography is widely known and admired. His images of the Maine coast, its islands and its working people are indelible and enduring. Colby College recognized this talent with an honorary degree in 2003. If the Island Institute has an image in the minds of strangers, it comes from Peter's photographs. But more than this, he is the Institute's Pied Piper, its cheerleader and fundraiser. He exudes the joy of the Maine coast on a sunny day.

Philip is a poet, an historian, a visionary, a wonderful and prolific writer and, in spite of all that, he is a very competent manager and leader of people. His knowledge of marine and natural sciences is encyclopedic. His knowledge of the Maine coast, its nooks and crannies, its ledges, its wildlife, the history and culture of its people is vast. He has an indomitable sense of humor and an intelligent and abiding love for the Maine coast and its island people. I simply cannot accurately do him justice.

Philip's and Peter's vision of Maine's islands and their communities is evident on every page of this book, as it has been through the twenty issues of *Island Journal* from which its contents were distilled. I sat up and took notice the first time I saw the *Journal*, and I can say truthfully that my attention has never wavered since.

Horace A. Hildreth Jr.
Chairman of the Island Institute's Board of Trustees

PETER RALSTON (2)

PHILIP CONKLING

THE IDEA FOR *ISLAND JOURNAL* EMERGED more than twenty years ago during an afternoon conversation with Betsy Wyeth, wife of the artist Andrew Wyeth. Peter Ralston, a freelance photographer, and I, a forester and writer, had been working on Allen Island for Betsy who had bought it in 1979 to prevent it from becoming a summer island of kingdom homes. Together we had developed an ambitious plan to turn Allen's 450 wild acres into a working farm, forest and waterfront.

By 1983 the collective efforts of Betsy, Peter, and a multifaceted crew under the direction of a Monhegan lobsterman, Dougie Boynton, had resulted in a radical, though controversial transformation of Allen Island. In a little less than three years, we had completed the construction of a wharf, launched a barge to haul pulpwood ashore, reclaimed fields, sown a pasture, introduced sheep, renovated dilapidated shacks, cleaned up abandoned wells, installed a portable sawmill, milled lumber, built a barn and made peace with most of the local fishermen. Each Allen Island project required rediscovering or reinventing techniques that had long since disappeared from the Maine islands as they depopulated, but were alive in the memories of islanders ashore and in stories and pictures we had heard or seen.

The people and stories that we had come across in reclaiming Allen Island gave us tremendous respect for both the complexities and rewards of island living. There seemed to be a need for a repository of knowledge about island living and a source of practical advice about how to manage islands that could occupy an ill-defined middle ground between the then-existing poles of seasonal residential development and "forever wild" preservation that represented the dominant uses of islands then. We discussed these ideas with George Putz, a visionary genius and writer from Vinalhaven, and Ray Leonard of the U.S. Forest Service who had funded an island research program I had worked on, and we began to hammer out the ideas for an organization that would become the Island Institute. The Institute would be a clearinghouse for information on Maine islands; it would provide forums for people with different points of view to discuss the islands' future; and it would sponsor island research and education. At the end of the summer, Peter and I went to Cushing to ask Betsy Wyeth what she thought of exporting the lessons we had learned on Allen to other parts of Maine's archipelago.

Betsy, the daughter of an accomplished newspaper man, immediately asked us about our strategy for persuading people that Maine islands were important. We had outlined a plan for a series of newsletters and monographs published by the Island Institute, but Betsy scoffed at that part of the plan. "No one will read those publications, except for a few people who already agree with you," she said. "You need to reach people who don't know or care about Maine islands," she told us. "And if you're going to try that, then *really* do it right," she said. We agreed to come back to her with a revamped publication strategy, and thus *Island Journal* was born.

Island Journal would be a high-quality, four-color, annual publication of the Island Institute to celebrate the Maine islands, profile islanders and cover the art and science of island life. Betsy and Andrew Wyeth agreed to help fund the first *Island Journal* by donating 10 signed high-quality reproductions of one of Andy's major paintings, *Reefer*, a haunting image of a man in a lighthouse tower. These reproductions, for which Betsy had pioneered a marketing and sales program, were selling for $3,000 apiece. Peter and I planned to offer *Reefer* for $7,500 to people we termed "Founding Members" in an organization that did not yet exist, except on paper, to fund the publication of the first *Island Journal*, which also did not exist except as a concept. We sent a mailing out to likely prospects, and sold two reproductions reasonably quickly.

Meanwhile, George Putz began contacting authors to submit stories and articles as Senior Editor of *Island Journal*. We persuaded islanders to write for a special section of the *Journal* called "Radio Waves" that would be a network of island correspondents (before reliable telephone service had reached islands such as Monhegan, Matinicus and Isle au Haut) to report on "news from offshore." We invited articles from the heads of the Maine Coast Heritage Trust, The Nature Conservancy and the Maine Seacoast Mission. Peter Ralston acted as *Island Journal*'s Photo Editor and Art Director; a position he has held ever since. Altogether we assembled the work of forty authors and correspondents and another twenty photographers and artists that we planned to publish in the first issue of *Island Journal*, with stories ranging from lighthouses, to fishing, to ferries, to island history, puffins, whales and eagles.

We hung up the Island Institute's shingle, under the welcoming umbrella of Hurricane Island Outward Bound School, an important island user. Peter Willauer, the farsighted head of the school, had convinced his trustees to support the overhead of the Island Institute's first year. I would raise the funds for my salary as Executive Director of the newly created Island Institute; the sale of the Wyeth reproductions would cover the production costs of *Island Journal*.

I was able to raise the organization's start-up funds during an unforgettable meeting with the late Tom Cabot in Boston. I had sent him a three-year Strategic Plan for the Island Institute and he invited me to visit him at Cabot Corporation's headquarters in Boston. I went prepared to pitch him hard. But before I had a chance to start, he called his secretary and asked her to bring in his checkbook, and calmly wrote out a check for the first six months' salary and operations. "Never be embarrassed to ask for money for something you really believe in," he said. Then he narrowed his eyes shrewdly and said that he would make an additional contribution the following year for a full year's salary, but then we must be prepared to make it on our own. He was as good as his word. I have always considered his advice about fundraising which can be a humbling, if not humuliating experience, to have been even more valuable than his electrifying start-up contribution.

Then reality hit. We received bids in the spring of 1984 for the printing of the first *Island Journal*. The only acceptable bid, given Betsy Wyeth's admonition to "really do it right," was north of $38,000. Our first year's budget was $60,000. We swallowed hard, as did Peter Willauer, who told us frankly that unless we sold more Wyeth reproductions and/or recruited more Founding Members, he could not approve printing *Island Journal* since his Trustees would not fund the shortfall. Stern reality, of course, helps focus the mind. Peter and I immediately went to work in the way each of us works best. I wrote pleading letters and Peter worked the telephone. Time was running out for the printing — it was now May 1, and without having the summer season to sell *Island Journal*, we knew we would be dead in the water. In desperation we finally identified two more Founding Members who we persuaded to contribute with the help of the offer of *Reefer*.

The last task before going to press was to choose the cover. Peter Ralston asked for help in editing his bulging photographic files, so Putz and I agreed to go through boxes and boxes of Peter's slides to hunt for a cover. The criteria we developed that day, we've adhered to ever since. The cover needed to be a strikingly beautiful island scene, of course, but it could not just be idyllic scenery. Since our purpose was and remains helping to visualize a balance between the hand of man and the hand of nature, we wanted people or a representation of human handiwork to be portrayed somewhere on the cover. When Putz found Ralston's photograph of the sheep in a dory towed behind a fishing boat toward Allen Island in the fog, we knew immediately we had our first cover.

Six weeks later, the shipment of the first issue of *Island Journal* arrived. It was the end of June; the Thursday before the beginning of the long Fourth of July weekend. Peter and I loaded 1,500 *Island*

Reefer, Andrew Wyeth, 1977

Journals in the back of his Jeep, and spent the next three days going first to the east and then to the west, peddling them door-to-door with a money-back guarantee. We stopped at bookstores, general stores, gift shops and any other likely place we happened upon. A big order was 10 or 20 copies; often we sold two or three on a trial basis. Anything to get them out there on the street. We ended the sales marathon at Portsmouth, New Hampshire, at the end of the day on July 3, with most of the *Island Journals* sold. We were in business.

For many years, we put together the story list for the upcoming issue of the *Journal* during winter meetings in a feverish charette, often on Vinalhaven, sometimes at Putz's sprawling farmhouse we called the 'Putzarosa,' where we ate and drank and argued over what kinds of stories we ought to write or find. Later we convened at Lane's Island in a 10-bedroom sea captain's house owned by my in-laws, the Morehouses. We sought out unique voices among the islands, including many unpublished

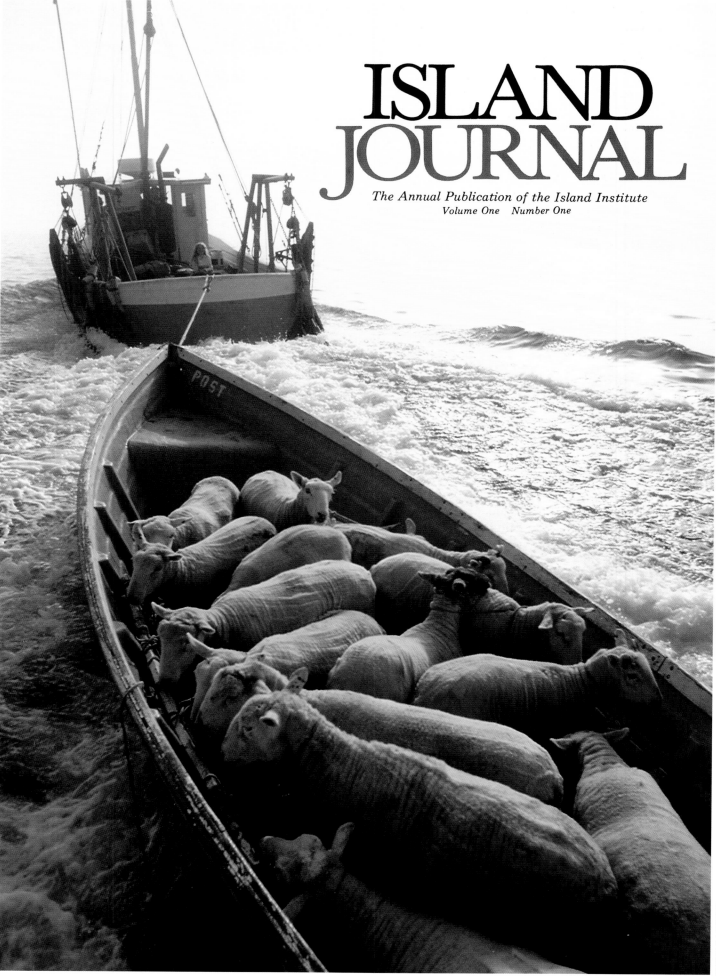

ISLAND
JOURNAL

The Annual Publication of the Island Institute
Volume One   Number One

PETER RALSTON

writers and photographers. We tried to find fresh ways of marrying words and images to create the kinds of circumspection and wholeness one finds over and over again in island experiences. We included art in the first issue and poetry in the third to add intensity to the collection of stories and voices, and to provide each issue with a spiritual grounding.

Mike Brown joined us for a while and then Cynthia Bourgeault, who was living on Swan's Island. She sent us some of her incandescent island writing, and we asked her to join the editorial team as our first managing editor. Michael Mahan Graphics in Bath has designed *Island Journal* since 1987. Ten years ago David Platt joined us and has managed with grace and apparent equanimity to make the *Island Journal* train run on time ever since. With David's genial professionalism and tireless persistence at the helm as editor, we now regularly receive more island stories, essays, poems, art and photography than we can handle. The well just keeps running over, a testament to the powerful hold Maine islands have on the imagination of the region; indeed, of the nation.

Perhaps the single most important element of the *Island Journal* has been the almost unlimited treasure of images we have been privileged to draw from Peter Ralston's photographic work. Many of Peter's images have become iconic; the sheep in the dory; Dick Lunt with his great-grandson Nate hauling a three-day-old cod by the tail; the sun dog over Old Man Island off Cutler — all have tapped into the collective unconscious of the Maine coast. Peter's greatest images remind us of archetypal experiences we've had — or would like to have had — while we are there at his elbow, finally seeing what his keen eye has already framed and his finger fired. Beyond that, Peter's photography has inspired the often-tired writer's mind in many of us writers to bestir ourselves to the enormous challenge of producing words worthy of his imagery.

Island Journal is only one thing that the Island Institute does. The rest of our work can be a long, hard slog, trying to be of measurable use to island communities who are rightly skeptical of off-islanders, especially well-meaning ones. At the Island Institute, we often say that *Island Journal* is the last fun thing we do. For many of our readers who come to the Maine islands infrequently, the annual issue of *Island Journal* is a way of reconnecting to the enduring sense of place Maine islands so effortlessly convey. It convinces them, once again, that islands and their communities need all the help they can get.

Twenty years ago, we piled up almost every chip we could lay our hands on and rolled the dice on the first issue of *Island Journal*. There was more concept than substance to the Island Institute then, and we bet the organizational farm on the publication. It paid off, not as a windfall, but in the long and slow process of gaining acceptance — a berth, if you will, among the proud and independent people of this amazing archipelago. And all of us who work on *Island Journal* today continue to be thrilled and privileged to be associated with this work in progress.

Philip W. Conkling

President, Island Institute

A NOTE ON THE PRESENT VOLUME: The editors went carefully through twenty volumes of *Island Journal* for the selections presented here. We have included stories and images from each annual issue that have stuck in our minds as the most memorable. Also, we have not simply re-published each story as it had originally appeared, but have looked at them in new ways. Some of the material has been trimmed a bit, in order to present a larger number of selections. Our goal remains the same as always: to pair words and images about the Maine islands and their communities that present their wholeness and their complexity in fresh and compelling ways.

HOLDING GROUND

THE BEST OF *ISLAND JOURNAL*

1984 – 2004

Chapter Four: Fish, Fishermen & Aquaculture

Chapter Five: Working Waterfronts

Chapter Six: Art of the Islands

Chapter Seven: Birds & Beasts

Chapter Eight: Islands Far and Gone

PETER RALSTON

CHAPTER ONE

Islandness

"ISLANDNESS" IS A CONSTRUCT OF THE MIND, a singular way of looking at the world. Articulating this perspective is perhaps more important to outsiders who for some reason associate themselves with islands than it is to islanders themselves, who understand the concept of islandness instinctively but may never feel called upon to express it in words, except for distinguishing between being "on island" or "off-island."

Islandness may be something experienced, like solitude; observed, like the ways islanders respond to change; or learned, like the lessons newcomers must absorb as they gradually become part of an island community.

Island Journal has explored islandness many times, most often through the eyes of people who "discover" islands and want to understand their meaning. "My definition of friendship and community has changed a lot over the last four years," writes Karen Roberts Jackson, reflecting on her own experience as an island transplant. "I feel sometimes as if my place is at the back of the line, behind a long list of people who have come before us. … Sometimes there seems to be no end to the dues that must be paid before the initiation into simply belonging."

British author John Fowles, a visitor to many islands, takes note of "the boundedness of the smaller island, encompassable in a glance, walkable in one day, that relates it to the human body closer than any other geographical conformation of land."

For Philip Conkling a defining moment was the solitude of a deserted island in eastern Maine, experiences there that "work like the tide and fog in strange ways of muffled sound and obscured sight to bring me back and back again to things that cannot be named."

Islands teach us lessons we forget at our peril, according to David Weale of Prince Edward Island. "For the island community, no less than for an individual," he writes, "the failure to respect the truth about ourselves is a serious and soul-destroying failure. Any repudiation of our Islandness is, therefore, a deep and fundamental repudiation of who we are — and of our uniquely precious existence."

Because it is a state of mind, islandness can be as simple or as complex as we wish it to be. But if we fail to appreciate it or don't pay attention to it — if we are guilty of "undersight," as George Putz would say — it will be lost, and the world will become a poorer place.

David D. Platt

Mosquito Island

The More Things Change
Why Recreational Culture Kills Island Life

George Putz

Older New Englanders will recall what Cape Cod was like 40 years ago, Martha's Vineyard and Nantucket 30 years ago, and the southern Maine beaches 20 years ago. All of these places have become very different from what they were. Among other things, these places used to be different kinds of places, unique places, not only with different kinds of geography and natural history, but also different kinds of heritage, uses of the English language, and patterns of interaction. By simple observation we understand that a pandemic kind of phenomenon happens because of recreational development. All these places are becoming more and more alike.

Our problem is not lack of data but lack of vision. I call it "undersight" — an inability to look up and see what is before us. Over the past 20 years, on my own island of Vinalhaven, we have seen the collapse of the "fish house culture" in favor of a more mercantile, recreational social structure. With an ever-larger number of supercilious and superfluous itinerant bodies infusing the landscape, the institutional fabric changes over time. I live in a place that has a work ethic — where you are a valuable human being if you're hammering with your hammer and sawing with your saw, or repairing a net; where you are busy. And if you're on the street, you're on the street to get something, on errands. You are occupied. As there begin to be larger numbers of physical bodies just wandering around, something in you gets a little rattled. I'm not talking about the pestiferous aspects of this; I'm talking about the symbolic ones. You know: "What the hell are these people doing here?"

On Vinalhaven we used to get away fine with one part-time local constable. Now, with no increase

Allen Island wharf

in year-round population, the island requires three officers — these because of the encroachment of essentially displaced itinerant personalities and their effects on the cultural fabric. As these new-comers increase, social requirements escalate by orders of magnitude.

You begin to lose poor people. As there come more and more itinerant bodies, more people ask about properties for sale. And they start buying them, placing demand pressure on the land. Land prices go up, pressure on the tax base increases, and the first to be displaced are the poor.

Then you also begin to lose the land-owning lower middle class, because they can no longer stand these burdens (or temptations) either. As this patrimonial alienation occurs, and you begin losing your poor and lower middle class working groups, this gradually saws away at the health of the community. In light of established mercantile values, this sounds like an odd thing to say, but the presence of these people *is* critical to community health. When they are gone, you don't have a community any more because you don't have a critical mass....

For all Vinalhaven's 172 miles of twisted, ledgy shoreline, there is only about a half mile of it, total, appropriate to sustain the wharves, fish houses, assorted piles of rope, traps, nets, and other necessary paraphernalia of the commercial fishing community — the working waterfront. When you begin to nibble away at even tiny bits of this truly useful shoreside land, you are literally cutting the ground out from beneath a community's traditional lifeways.

Just as soon as you see an interest in convert-ing to recreational uses, the game is basically already over, even for the highline fishermen, the quarter-million-dollar-cash-turnover guys who can afford their required work spaces. Beyond the costs of wharfage and fish houses, places to stack gear, and so on, when the waterfront market is tuned to recreation, you get more and more people who don't like the smell of bait, piles of encrusted rope, and guys who say things like "fuck" in loud voices, all day and night, sometimes little else. Ostensibly genteel people don't like to hear that.

And eventually the new genteel sensibilities will prevail by sheer economic clout. Take bank-ing, for example. Banking practice is one of the first things to change when a community's eco-nomic focus shifts. If a loan officer in an island branch bank has a choice to make between an apparently iffy $15,000 loan for electronic fishing equipment, using the boat as collateral, and the same amount of loan toward a terrestrial recre-ational establishment, you can forget the electron-ics. Risk capital is just like any other addictive predication — you go with the best, least risky option available.

Expectations shift. The community becomes "dispirited" as it shifts from the Masonic order, the Red Men, or other sorts of secret societies and religious groups, where there's a direct connection to the mysticism of the family, of what it means to be a man or woman, wherein life itself is fused with mystery. All this gets pared away, until you have a much more simple-minded, pragmatic, symbolically impoverished institutional life.

And this dispiritedness, sadly, works its way into the next generation. For 200 years of island history, a 12-year-old boy, if he was going to amount to anything, was going to be a fisherman, a boatbuilder, or in some other way a principal in the maritime community.

There are no more roller skates on the street or sidewalk. There is no bike with a broken sprocket, and there isn't any laundry flapping in the breeze or gossip going back and forth between the hous-es. It's just a dead place, with the wind whistling through the eaves. And those kids that aren't there anymore aren't in school becoming distinct, proud islanders. We've not only changed history in this regard; we've in a large part eliminated it. Through undersight we have allowed yet another vibrant part of cultural diversity to slip away from our human family.

1993

JEFF DWORSKY

THE ISLAND

Since I'm island-born home's as precise
as if a mumbly old carpenter,
shoulder-straps crossed wrong,
laid it out,
refigured to the last three-eighths of shingle.

Nowhere that plough-cut worms
Heal themselves in red loam;
Spruces squat, skirts in sand;
Or the stones of a river rattle its dark
Tunnel under the elms,
Is there a spot not measured by hands;
No direction I couldn't walk
To the wave-lined edge of home.

In the fanged jaws of the Gulf,
A red tongue.
Indians say a musical God
Took up his brush and painted it;
Named it, in His own language,
"The Island."

Milton Achorn, *1991*

Islandness

David Weale

Iwas driving with my 10-year-old son along the shore on the way to a late afternoon hockey game in a town an hour or so away. He sat quietly in the seat just looking out the window at the passing landscape and seascape. I turned to look at him several times, but he didn't even notice. He was absorbed in his looking. Then it occurred to me what he was doing. He was taking in the landscape. He was, if you will, ingesting the Island. And that is exactly what happens when you live here for long — you take the Island inside, deep inside. You become an Islander, which is to say, a creature of the Island. Islandness becomes a part of your being, a part as deep as marrow, and as natural and unselfconscious as breathing....

Wherever we look in the world we discover peoples whose lives and cultures have been shaped by their natural environment. There are mountain people, valley people, and people of the open plains. There are polar people, coastal people, and people of the forests. In each case the nature of the community — its mythology, imagination, its very soul — has been sculpted and colored by its geographical

PETER RALSTON

circumstances. Further more, it would certainly be foolish to think of any one of these as being superior to the others. Each is good and bad in its own way, and its strength and genius is derived from its adaptation and response to its own geographic peculiarities....

The uniqueness of an Island is its geographic precision. "An Island," as I was taught in grade four, "is a body of land completely surrounded by water," — or, as a friend from Cape Breton recently expressed it, "An Island is geographically perfect."

The topography and landscape of this province — that is to say, its Islandness — is the source and reference point for the imagination of Islanders. It is the primal source of our communal insight and wisdom....

For the island community, no less than for an individual, the failure to respect the truth about ourselves is a serious and soul-destroying failure. Any repudiation of our Islandness is, therefore, a deep and fundamental repudiation of who we are — and of our uniquely precious existence.

The term Islandness, or "insularity," frightens some people. They think it is a sign of narrowness or narcissism. There is, however, no inherent contradiction between "Island-identity" and "global citizenship," any more than there is a conflict between the healthy self-esteem of an individual and her ability to participate in community life. Indeed, the two are not conflictual, but complementary. In other words, my identity as a citizen of this Island we call Prince Edward is a complement to my identity as a citizen of this Island we call the Earth...

1991

Crazy on the Rock

Karen Roberts Jackson

For a brief period in my life I lived on the Hawaiian island of Oahu — six weeks at the most, just past my twenty-first birthday. Still, I refer to that time as having "lived" there because I had transferred across the ocean all my meager worldly treasures and my wildest dreams. Within a week of my arrival, the boxes bearing my possessions were delivered by my Polynesian post-mistress with a meek smile. Each box tinkled like tree branches after an ice storm; the contents looked like the multicolored shards inside a kaleidoscope. As I peeled off the duct tape from each crumpled carton, I could sense the fracturing of my lofty dreams as well.

I had followed love to this paradise, but even love could not sustain me. While there were passion fruits and strawberry guavas offering themselves up in this Eden, there were no jobs to be had to pay the rent. I was not native, nor could I speak Japanese: just another poor, white-faced parasite, a *haoli*, as the

PETER BENCHLEY

locals say. In my backyard were lush, green mountains; in my front yard was an endless expanse of the most exquisite blue ocean. But somehow all I was conscious of was the singular King Kamehameha Highway that encircled the island. I had spent every dime of my waitressing career flying into Oahu airport. I felt a strangulating kind of claustrophobia knowing I could not get off the island.

The locals had a name for the haoli sickness: they called it "Rock Crazy," and in my naive youth, I probably caught the malady quicker than most. I missed friends and community and was not prepared for outward hostility. I was too young to know that a sense of belonging takes time and nourishment.

So it is ironic, I suppose, that I write this piece now from another Rock, in the middle of Penobscot Bay, an hour and a quarter ferry ride from the mainland. It is three days after Christmas; the temperature has hung at zero all day. My feet are propped on the chrome bar of the woodstove, and I write wearing wool fingerless gloves. My breathing creates puffs of smoke that steam my glasses.

This will be our fourth winter here, and the season is just beginning to show her stuff. Fool that I am, the dread of winter consumed half my summer and stole nearly all of the splendor of autumn. In a sense, it is a relief for the frigidness and the inherent "bucking up" to be here at last. I will admit with all candor that I have been battling the Rock Craziness for many months now. I have tried short trips away, rearranging the furniture, and working myself so hard I collapse at the end of the day. These methods all help in short spurts, but I know that once again the core of the matter is isolation, the lack of friends who can relate to my daily life, and a hunger to have a sense of belonging to a community.

When we moved here I was again following that love thing, this time in an expanded equation that included a family. My partner and I counted our wealth in increments: four children, four acres, lots of lumber, and endless dreams. Older and supposedly wiser, I choked back my misgivings as friends transported boxes of soggy books (my precious cookbooks) up from the shore and promised to visit the following summer. It would not be the last time that I felt like a pioneer family breaking loose from the wagon train. We rolled up our sleeves, dug in our heels, and got to work creating a home. Perhaps it was the seemingly endless labor and lists of things to do that kept our minds plenty occupied for a while. But now that we have a roof over our heads and a fleet of boats to serve every purpose, I seem to have more time for introspection. The question I find myself chewing on the most is: how do my fellow islanders keep from going nutskie? And tell me again why we chose this incredibly complex "simple" life.

A quote hanging on my kitchen wall, from the Findhorn community in Scotland, reads: "Become more aware of the things in life that really matter; those that gladden the heart, refresh the spirit, and lift the consciousness…." Depending on the day, I am not sure whether this is my creed or my curse. What refreshes my spirit is laughter and kinship, commodities found only in the presence of other beings. Despite the stark beauty that surrounds me — the white woods, ice water, gray granite — it is all I can do to fight the chill in my heart and the bone-loneliness.

Against this backdrop, my definition of friendship and community has changed a lot over the last four years. I feel sometimes as if my place is at the back of the line, behind a long list of people who have come before us. I am not a native Mainer, although two of my children were born in Maine. I am not a native islander even though I all but sleep in my mud boots. Even on our small island, made up of a handful of families, we are neither the first to come nor the longest to endure. Sometimes there seems to be no end to the dues that must be paid before the initiation into simply belonging.

I've also been here long enough to feel much sympathy for why it is that this ritual exists. Friends — and more and more, strangers, out of the blue — come to visit us here, in the summer, of course. I wince when their first words of greeting are, "Got any land for sale?" They bring guests of their own to show off this paradise; they pontificate like tour guides on our "alternative lifestyle," our "intentional community." Increasingly my response is, "Come check out our alternative lifestyle in February…."

Yet in my heart, I do not mean to be rude or withdrawn. The first visitors of summer are often my lifeblood. They bring news and tidbits from an outside world that I sometimes forget exists. I find, however, the longer I am here, that I must overcome a shyness around visitors that surprises even me. Either I don't know what to say, or I can't shut up. The mental agility and social graces that would normally enhance such an exchange seem to trip all over one another.

Last summer a couple came to visit bearing wondrous gifts. Wade has been coming to his family's little cabin on the island since he was a child. He is now a prestigious artist in Philadelphia, and his sweetheart, Kate, is a poet and museum curator. On the way here, they had stopped in a Chinese market and brought my children such delights as Chinese yo-yos, rice candies with edible wrappers, soaps, teas, and incense in exotic paper. For me, Wade had brought a book, one of his paintings comprising the jacket cover. It was a collection of contemporary literature, new works from some of today's brightest and most promising writers. The gift nearly brought me to tears. I stood in a corner of my kitchen awash with emotion. It was as if Marco Polo himself had landed on our shore, reminding us that indeed the world is round and chockfull of exotic smells, colors, thoughts, words. What's more, I was reminded that I exist as a particle of that planet.

When my manners returned, I thanked them for their gifts and asked what had prompted such generosity. Wade's sincere response was, "Hey, we live in Philadelphia, you guys are the closest neighbors we have…."

In no way do I mean to diminish the soul sustenance provided by my island neighbors; both on my own island and on neighboring islands. I heard somewhere once about a contraption, a bosun's chair sort of thing, by which a pair of schooners could draw near to one another and the captain's wives could swing between the two ships for much needed chitchat. Or maybe I just dreamed it, as I often daydream about some convenient way to stop in for a lengthy cup of tea in a woman's warm kitchen. Instead, I seem to engage in stoic exchanges about the weather, brief comments in the post office, and try to build on these acquaintances over time. Always, I feel, there is the persona of toughness to uphold, the rusticator. A few months ago I wrote a piece for a literary magazine called *Feast*, produced each summer by a summer resident. I happened to visit some of my island neighbors one day as they were perusing the magazine. With all good intent they were joking that a winter counterpart should be produced, entitled *Famine*, on ratty newsprint, that told the true story of island living. I laughed along at the idea, but inwardly I cringed. Is it really necessary to constantly remind ourselves of the harshness of our life? Is adversity our strongest common thread?

This past summer I stumbled into the Farnsworth Art Gallery, sharing time with a visiting friend until the next ferry run. In addition to the regular Wyeth family exhibit on display, there was a special exhibit of Jamie Wyeth's spectacular work. One painting in particular, *Breakfast at Sea*, struck me as if the wind had been knocked out of me. The painting is of an idyllic breakfast for two, on a long white porch overlooking the sea and the other islands. The tension between the couple is obvious. The caption read: "Everybody started looking at their relationships, and saying, 'Do I really want to live on this rock for the rest of my life?' and a lot of divorces happened…. The older islanders were quite anxious because you need young people to support an island, to give it stability." The statement that other island couples had questioned their own ability to persevere, their own stubbornness, was a revelation to me. I had

believed that confessing one's need for something more than nature's beauty and backbreaking work was equal to an admission of defeat.

The statement, too, that the older folks kept a watch on such things, made me smile. I recalled the anxiety I felt, when first I came to the islands, of keeping names straight, who was related to whom, and how that might affect my life. I remember thinking that the scientific discoveries of body language and subliminal messages were nothing new to island folks. They had generations ago learned the art of observation, of reading facial expressions, tone of voice, the handwriting on the wall. It made me nervous, as if the community knew you were bound to screw up and that it was just a matter of time before you cracked.

But there is a comfort as well, in this sense of people keeping watch over each other. While my hunger for long soul-sharing conversations may go unfulfilled, I seem to have found it easier to have intimate relationships with people I hardly know, and often, have never seen. Our main source of communication with the outside world is via the VHF radio. Whatever you say is heard by all who have their set on. I adore the fisherman who consistently calls his wife "dear" and always lets her know when he is about to pass by the house. The mates who check in throughout the day, who

Breakfast at Sea, Jamie Wyeth, 1984

describe in detail the warm vittles that await on the table at home. Like an audio soap opera you learn to tell by an unspoken code who is feeling under the weather, whose wife just went into labor, who has frozen pipes needing attention at home.

When we first moved to the island it was a daily game show figuring out who was taking the kids across to the larger island to school and picking them up in the afternoon. One day I was in a coffee shop in town and a silver-haired, sparkly-eyed woman said to me, "I know you; you're Rosebud." I laughed, a bit startled, and replied "How did you know?" She said, "Oh, I hear you each morning on the radio." I said, "Well, if you ever have any suggestions for us, just jump right in." She replied, "Oh, I don't talk, I just listen...."

I realize that this familiarity is still a bit alien to me, not a part of my gypsy upbringing. I come from a background of seekers and dreamers, "lost souls" as some would call us. As nomads often do, my parents sought warmer climates and settled for a time (my childhood) in places such as Orange Blossom Estates, in Florida, subdivisions with street names like Ponce de Leon, Hibiscus, and Zephyr. I have been to my hometown once in the last ten years, and only once in the ten years before that. There was no old neighborhood to cruise, not a familiar face on a sidewalk; the town is virtually unrecognizable from my youth.

My maiden name is Roberts and once I stood in a cemetery on Lane's Island, startled to find myself in the company of several Roberts there. I fantasized that perhaps I did have roots here, that one of my ancestors might have been a sea captain (I was told as a little girl that my South Carolina relations had been sailing men). I imagined an earlier time when the islands were first inhabited by explorers, immigrants, dreamers such as I. A time when island folks' homes faced the ocean, the watery highway that had brought them there, when their homes contained carpets, tapestries, and souvenirs from distant lands.

My husband, on the other hand, grew up in a medium-sized Midwestern farming town. To this very day he can go home and be greeted on the street by passersby. He is remembered for his athletic victories, his club-related awards, and his boyhood mischief. More important perhaps, he is reminded of his family lineage; his father's and grandfather's boyhood mischief and adult achievements.

It has taken this extreme introspection of late for me to question what it is that we hope to give our children here. The long winters alone have required us all to become friends as well as family, something unheard of in my childhood. My immediate island neighbors treat the children respectfully and lovingly, giving them an extended family of aunts and uncles, and mentors when the need arises. I realize I hope to give them a blend of both, a healthy strand of gypsy blood and yet a strong sense of homeland.

For myself, I am still defining the balance.

1994

PETER RALSTON (2)

THE WOODS AFTER RAIN

The damp trail through the woods
Becomes more lost each year,
Runs over the rocks and roots,
Through waves of deep-piled moss,
A static sea,
Land-roll of the forest floor.

Here once for a loaf of bread,
I took an enchanted way
Through path turn pebbled stream,
And brought back shoes
Wet with the overflow
Of twig and berry, rock-drip,
Some touch of sea,
Or ooze of blackened pools
Where leaf and dead ferns rot.

Darkness held off,
But silence closed me round,
For birds had hidden away;
The crowding trees, vivid with lichen's jade
On ink black trunks,
Made a green gloom,
A tunnel for a ghost.

I might have felt alone,
Yet walked companioned, friended,
As if I, too, shared the rough mothering
Of island earth,
Fed as a root on scanty nourishment,
And drew from her wild marriage with salt
wind
Close kinship, strange adoption—
My own foot-fall
Now native here.

Hortense Flexner, *1990*

Solitude

Philip Conkling

I first set foot on the Maine islands 16 years ago. A graduate student at the time, I was working toward a degree in forestry and I wanted more than anything to go to work somewhere within the vast reaches of the North Maine Woods: the place, I thought, where timber cruising foresters and other real Maine men went. The job market was not encouraging of my fantasy, but on a bulletin board at the forestry school I happened to see a mimeographed notice listing Earth Day-type environmental internships in the Northeast that included one in Maine collecting natural resource data on 12

PETER RALSTON

islands owned by a nature group. Like all truly ignorant people, I had some preconceived ideas about Maine islands because, after all (I thought to myself), I had seen some of them from the peninsulas of Steuben, the small Washington County fishing and blueberry village where I had been living for the two previous years.

Steuben's narrow peninsula roads wandered far down and away from the well-traveled Route 1 — that lifeline that took us either east or west to the big shopping towns of Machias or Ellsworth. My favorite byway, the north-south trending Pigeon Hill Road, ran down to the tip of Petit Manan Point, which splits Narraguagus and Dyer Bays. Off to the east from Petit Manan, the view encompassed a dozen or so mostly small, spiny, porcupine-like islands crawling down the bay on their way out to sea.

I thought I knew two things about Maine islands when I started to work on them that May of 1975. I knew they were rocky and I knew they were covered with spruce. I also had a third suspi-

PETER RALSTON

cion that I didn't voice while I was preparing for the field season; namely, if you had seen one of these islands, you had seen them all — or certainly most of what needed to be seen.

I suspect I had gotten the job of surveying the natural resources of the 12 islands I was to visit mostly on the strength of having known a few Washington County fishermen who were willing, in a loose sort of way, to help out with transportation. Eager to get under way with my assignment, I packed my kit for a three-day trip to Flint Island, a 100-acre island at the outer edge of Narraguagus and Pleasant Bays.

I left aboard the lobsterboat JESSE from Pigeon Hill lobster pound, tucked up inside Petit Manan Point, in a dungeon thickness of fog. In those days radar was still a very expensive item for inshore fishing boats, and so we steamed out on a compass course past Dan Leighton Point, north of Pond Island, then four miles across outer Narraguagus Bay headed for Flint — or more precisely, as I was shortly to learn, for the bell buoy near the entrance to the Flint Island Narrows. Not until many years later would it have occurred to me that we were beam-to in a big tide and seaway where two large bays meet the waters of the North Atlantic and where eastern Maine's normal 12- to 14-foot tidal range pushes its relentless way around the islands. Nor did it occur to me that our course, which skirted the shoals north of Pond

Island and Western Reef off Shipstern Island, might have presented any special difficulties.

My lobsterman friend said almost nothing for the first 20 or 25 minutes after leaving the lobster pound. Then he throttled back, cut the engine, and stepped to the stern of the boat — "Listenin' for the bell," he said. But there was only the muzzy, muffled sound of the sea's rote, and after a minute or so he fired the engine back up, steamed another few minutes, shut down, and listened again. He listened hard, as swells slapped at the sides of the JESSE. Slowly I began trying to separate out sounds, all strange, dampened down in the fog, coming from all directions at once, with no horizon to fix on, with no direction known. As the seconds ticked by, I heard nothing faintly resembling a bell. Nothing like, whatever, maybe a gull call careening off somewhere.

"There," my friend said finally. Taking a quick course, before he lost whatever it was he heard, he throttled the JESSE ahead again until we stopped for the third time. By now I, too, could hear the chaotic clap of the bell buoy, as I tried to figure out where the sound was coming from out of the dense fog. Then the lobsterman pointed into the thick white blankness, "There it is." Though I stared straight and hard directly to where he was pointing, I saw nothing. I'm not sure I really believed he saw something, but I kept staring to the place he pointed as he throttled back up. In another 10 seconds we motored by the clanging bell and three minutes later we steamed into the little anchorage of Flint Island.

I distinctly recall these few thoughts: At first, I was disconcerted that this lobsterman could hear things I couldn't hear and see things I couldn't see. And then slowly it began to occur to me that perhaps there was more out here than met the eye. Soon I was ashore by myself. After pitching a tent, arranging my kit, laying out field guides, botanical keys, binoculars, hand lens, and such, I fell to the task of taking stock of everything around, writing down long lists of names of things I found — *Lathyrus japonica; Picea rubra; Mertensia maratima; Iris hookeri; Empetrum nigrum*, and so on. The names rang out over the next few days as I wandered the shores and swales and beaches of this uninhabited island, absorbed in my cataloging.

I was struck by the intricately beautiful pattern of the white cherty cliffs for which Flint was named, and although it took several more months for me to fully realize it, I had landed on an island like no other on the Maine coast.

The interior of Flint, like that of many islands I was to visit that summer, was a dense thicket of tangled young spruce and alder hells, best negotiated on one's hands and knees and completely discouraging to anyone not being paid specifically to transect them. I found in Flint's interior only one entirely arresting feature that was to concern me on and off for most of the rest of the summer. Within the densest interior of this lonely island were stone walls and piles of fieldstone. It wasn't until much later that I was directed to the home of one of the Sawyers in Milbridge, an old man who had tended a flock of sheep on Flint Island as a boy and told me who had lived there and what had become of the people of the island. But on this particular foggy day, I began to realize that this island, like the people around it, held secrets that were not likely to be deciphered by the small library I had brought along.

Amid these few disconcerting thoughts, I spent my allotted three days alone and had acquired a long list of names of what grew on the shores, flew through the trees, and lived in the rocky intertidal zone. On the morning I was to be picked up by my lobsterman friend, I was feeling pleased with myself and was all packed and ready to go, down by the shore. I recall I waited a couple of hours, until the first wisps of doubt crossed my mind that maybe the JESSE wasn't coming back. Although the middle day of my visit had been bright, blue and sunny, the fog had now returned and had probably delayed my lobsterman friend who, I was certain, couldn't have confused the days and wouldn't have forgotten me. By this time I had pretty much eaten everything I had packed out. Along about mid-afternoon, I had concluded I wouldn't be seeing the JESSE come out of the bleak whiteness off Flint's northwest point, so I re-pitched the tent, went over my notes, added to my lists, and tried not to think about eating.

The next morning I woke up early, but decided against packing up my gear — I suppose as some kind of propitiation to the events of the past day. I began walking down the flinty, shingly shore over which spiky spruce boughs were combing tiny droplets of water from the wet breath of fog. At such times everything is very close, the view is narrowed to 40 or 50 feet, with little to be seen, and because not much is moving on the water, the air is still, heavy and silent.

At that moment, directly in front of me, not 20 feet above the beach and just off the island edge, navigating by the shore spruce, I was staring at two magnificent bald eagles: one a female, slightly larger than the male. They flew by, wingtip to wingtip, out of the fog — as startled to see me as I them, and then careened sharply away and were gone. I heard, and in the damp air imagined I felt, the rush of heavy air from their wings on my face. Although I'm not sure I knew it then, in that moment my life changed forever. In that experience that lasted an instant, I felt a sensation telescoping itself outward, beyond Flint's outermost realm, into a foggy white light far beyond boat times and lists of names of things I had left at the campsite.

Over the past 16 years, I can think of another small handful of such experiences, which almost always occur alone and accidentally; which strike me dumb at the time, but then work like the tide and fog in strange ways of muffled sound and obscured sight to bring me back and back again to things that cannot be named.

Here in Maine, still within our arms' reach in the late 20th century, we find a multitude of these once lightly inhabited islands, places that are rich in the ways of maritime history and culture. Places that, the smaller and more enclosed they are, the larger the window on the infinite, the farther they telescope to the heavens.

These are spirited and peopled places, and we must consider carefully how to keep these worlds balanced between accessibility and inaccessibility — because in one single moment of solitude, they provide our callous, name-collecting natures something as precious as vision itself.

1991

Siren Call

John Fowles

True islands always play the sirens' (and book-makers') trick: they lure by challenging, by daring. Somewhere on them one will become Crusoe again, one will discover something: the iron-bound chest, the jackpot, the outside chance. The Greek island I lived on in the early 1950s, Spetses, was just such a place. Like Crusoe, I never knew who I really was, what I lacked (what the psychoanalytical theorists of artistic making call the "creative gap"), until I had wandered in its solitudes and emptinesses. Eventually it let me feel it was mine: which is the other great siren charm of islands — that they will not belong to any legal owner, but offer to become a part of all who tread and love them. One's property by deed they may never be; but man long ago discovered, had to discover, that that is not the only way to possess territory.

It is this aspect of islands that particularly interests me: how deeply they can haunt and form the personal as well as the public imagination. This power comes primarily, I believe, from a vague yet immediate sense of identity. In terms of consciousness, and self-consciousness, every individual human is an island, in spite of Donne's famous preaching to the contrary. It is the boundedness of the smaller island, encompassable in a glance, walkable in one day, that relates it to the human body closer than any other geographical conformation of land. It is also the contrast between what

Above: *Bronze Age,* Jamie Wyeth, 1967

PETER RALSTON

can be seen at once and what remains, beyond the shore that faces us, hidden. Even to ourselves we are the same, half superficial and obvious, and half concealed, labyrinthine, fascinating to explore. Then there is the enisling sea, our evolutionary amniotic fluid, the element in which we too were once enwombed, from which our own antediluvian line rose into the light and air. There is the marked individuality of islands, which we should like to think corresponds with our own; their obstinate separatedness of character, even when they lie in archipelagos.

Island communities are the original alternative societies. That is why mainlanders envy them. Of their nature they break down the multiple alienations of industrial and suburban man. Some vision of Utopian belonging, of social blessedness, of an independence based on cooperation, haunts them all.

Since the proximity of the sea melts so much

PETER RALSTON

The Wind, Jamie Wyeth, 1999

in us, the island is doubly liberating. It is this that explains why indigenous small island communities, at least in the long-discovered temperate zones, are on the whole rather dour and puritanical in their social ways and codes. They have to protect themselves against the perennial temptation of the island: to drop the necessary inhibitions of mainland society. Islands are also secret places, where the unconscious grows conscious, where possibilities mushroom, where imagination never rests. All isolation, as the cold bath merchants also knew, is erotic. Crusoes, unless their natures run that way, do not really hope for Man Fridays; and islands pour stronger wine of forgetfulness of all that lies beyond the horizon than any other places. "Back there" becomes a dream, more a hypothesis than a reality; and many of its rituals and behaviours can seem very rapidly to be no more than devices to keep the hell of the stale, sealess, teeming suburb and city tolerable.

All desert islands, perhaps all desert places, are inherently erotic, as countless stranded individuals have realized. We all, whatever sex we are, want to know why we are alone; why has that universal yet obstinate human myth, the one we gen-

erally call God, so forsaken us?

I first began to feel the releasing power of *The Tempest* when I lived on Spetses ... the lack of a Prospero, the need of a Prospero, the desire to play Daedalus. It is the first guidebook anyone should take who is to be an islander; or since we are all islanders of a kind, perhaps the first guidebook, at least to the self-inquiring. More and more we lose the ability to think as poets think, across frontiers and consecrated limits. More and more we think — or are brainwashed into thinking — in terms of verifiable facts, like money, time, personal pleasure, established knowledge. One reason I love islands so much is that of their nature they question such lack of imagination; that properly experienced, they make us stop and think a little: why am I here, what am I about, what is it all about, what has gone wrong?

1999

Witness to the Deep

David Conover

One day, over a decade after lift-off, the VOYAGER robotic spacecraft reached a point on its outward journey where visual contact with the earth was about to be lost. For no scientific reason, astronomers turned its cameras back and focused them towards home. There in the distance was a tiny warbling speck of pale blue light. Another few hundred miles of travel and the blue light disappeared. Experiencing the threshold between a known light and the deep has been no less mysterious and no less compelling for generations of mariners and lighthouse keepers. At this threshold, a lighthouse guides the voyager with a tangible comfort that no radar, no radio, no GPS, no outer space navigational system will ever replace. The connection is primordial, residue of a nomadic, campfire-centered past. In the darkness, actual visual contact with a known light tells us where we are. In the darker darkness, it tells us that we are.

Five years ago, a Norwegian friend and I were groping our way ashore in a small sailboat, having been dismasted towards the end of an arduous transatlantic passage from Europe. As navigator, I hadn't been able to get a celestial sight in nine days, but rough calculations and our electronic instruments put us 30 miles off the Cape Race Light of Newfoundland, an island landfall light that was the first sight of North America for millions of immigrants. We were closing in on the rocky cliffs. Night approached. We wanted to see land, to see a light, to know where we stood. Suddenly, our electronics told us we were in Winnipeg, thousands of miles away.

DAVID CONOVER (2)

PETER RALSTON

Searching for the light that night was one of the many times I have understood the significance of a light for a mariner. If you have not been to sea at night, imagine this. You're confused, possibly lost. Darkness distorts the strength of the wind on your face and the size of the waves beneath your hull, making everything far away seem ever so close. "Where is that light?" you ask as you strain to see. Your look into the dark during these moments is absolutely sincere, absolutely focused, like the reaching of the shipwrecked for that first piece of floating debris, because finding the light may mean your salvation.

Some years later, I had the opportunity to meet the man who had been the keeper of the Cape Race Light that night, safely perched high on the cliff. Fred Osborne had been a keeper for over 30 years, following the course of his father and grandfather. Cape Race was his last assignment, one of the few remaining manned stations on the entire eastern seaboard of North America. He didn't have much to do, except keep the place spotlessly clean and occasionally wander up the winding steps of the 96-foot tower. As a hobby, Fred kept a ham radio watch with the call sign VO1JO. The keeper at Cape Race, he told me, was the first land station to receive the SOS of the sinking TITANIC.

Back then, mariners depended on more than the beacon being there. Like the pale blue light of distant island earth, all lights were lived in, kept, not remotely controlled from afar. A tower was built under the light and houses were built next to towers. A keeper was always nearby. The light was a home, an extension of domestic architecture, another odd-shaped room in a simple house where brass was polished, bread baked, cows milked and a watch kept. Unlike navigation lights today, the lighthouse station was designed not only to be seen, but also to see. For the keeper, the light tower was like a giant optical instrument, a watch-tower during the day, a magnificent lens and articulated light at night. Eyes and observation mattered in this place, a visual place. This fact often goes unappreciated because most people know light stations from afar, looking at the light from a boat at sea or a car on shore rather than with it like Fred Osborne.

In Maine, my family have been seasonal residents of a former lighthouse for over 15 years. Our caretaking responsibilities do not include the fixed green light itself, only cosmetic work on the tower. The Coast Guard stops by every six months for the light, which is still active and automated. Consequently, I've always felt a visitor to the tower, apart from it and its original purpose. I've often wondered how the task of tending the light affected the lives of the keepers, what they thought about, imagined. I've read first-person accounts of keepers' lives, looked through the extensive amount of lighthouse literature and images available in the coastal bookstores and museums, but still find myself returning to the structure itself, viewed up close.

I imagine a keeper's trip into the tower late one night sometime in the 1800s, perhaps to fill or pressurize the kerosene lamp. This task had to be done every four to six hours, around the clock. The heavy door is unbolted and swings open with a creak. The tower is dark inside, still and attic-like, with a smell of machinery oil and smoke. Sounds bounce off the circular walls and echoes layer on echoes as one ascends the curved stairs. At a landing, a crack of light from the ceiling trapdoor shines on a ladder. When the trapdoor is opened, light floods downward. The keeper eventually reaches the top room, checks the flame, then adjusts the wall vents to match the wind change and keep the correct draft. The lens and the blue glass around the lamp need dusting, but that is nothing compared to the outside windows covered with moths and bugs attracted to the biggest night light around.

There is a small door to the catwalk outside, where the wind and the sound of the surf can be heard below. Outside, a few steps away, the keeper could look back. The light is green, a color that results from the yellow beam of a lamp passing through blue glass. Turning towards the darkness, the light appears reddish as the cones and rods in the eyes struggle to compensate. The green beam hangs in the faint night mist, focused by the Fresnel lens to a 15-degree arc that covers all the water except for the surf below. A light would only reveal the danger of that place.

The mind wanders for a moment, following the beam projected out. Down in the distant darkness, a passing sail of a coastal schooner provides a tiny white screen for the light to play out its colored ID. Faint voices carry with the wind. The keeper's eye sees shadows — never noticed those before — black arms radiating out, one arm for each of the catwalk's railings and each of the window supports for the glass. Walking around the tower catwalk, the keeper sees the white house next door is also bathed in green. I wonder if the keeper's hands would ever make a shadow rabbit on its wall before heading down to ring the bell. That rabbit would be two stories high.

Keepers of the 1800s were communicators in a different age, before the telegraph, the wireless, the telephone, and the age of information. Back then, communication was tied to the limits of the human body and the pace and direction of its movements from place to place. If you wanted attention you shouted, blew a horn or waved a flag. Letters were hand-delivered, carried by someone who walked or rode a horse or a train. News was understood in context, usually local and relevant. Hardly ever was it trivial. People paid attention to it, because it was accompanied by other people. Every message had a messenger. A lighthouse keeper was an extraordinary messenger with a singular goal: keep the light lit. To this end, each keeper was dedicated.

Along with the light stations, the keepers have attracted a good deal of attention from those romantics in love with the sea. Solitude, adventure and a unique positioning at the shoreline have everything to do with this. The light stations, whether they be coastal on a mainland peninsula, among the islands, or further out still on a rock, were always the gateway through which dreams of escape to the open sea could pass. As the energy of these dreams passed on to the keepers themselves, the question arises: Of what must those gatekeepers have been made? How did they remain calm in the storm, prepared for being cut off from supplies, with the imperative of keeping the light burning organizing their life? Eternal vigilance is the price of safe navigation, and the keepers paid the price.

Born in an age when lights were powered by the flames of a whale oil lamp and ships were powered by the wind, these aids to navigation marked a coastal waterway that was the only Interstate of the day. By the late 1970s, however, the situation in almost all the Maine stations had changed. Keepers packed their trunks with tools, paint and polish and walked down to the boats, leaving their towers and houses behind like outgrown shells. These shells had capacity, but no longer any purpose. The age of sail had passed. Most coastal traffic had moved ashore. Lights became powered by a generator or by the mainland electrical grid. Initially, the keepers' houses were boarded up and minimally maintained by Coast Guard commanders with a strong nostalgic attachment to the stations. Gradually, a few became used as research stations, museums or parks. Many years into the future, an archaeologist uncovers a primitive structure along a shoreside cliff. A signaling cairn of some sort, a few thousand years after the Druids and Stonehenge. A Promethean attempt to bring the fire of the stars to earth. A time when the ocean was important and central to people's lives. What will that archaeologist understand then, and what understanding will he have lost? Attention is then turned skyward to remote travel to distant planets, where stars are once again navigational aids. Looking back, the blue light remains. The magic of the light station is understood, perhaps, forever.

1995

CHAPTER TWO

Islanders

WE HAVE EDITED AND WRITTEN FOR ISLAND PUBLICATIONS for quite a few years now, and we have learned one thing: islanders aren't particularly interested in being "described." In fact, we suspect that they're sick to death of being analyzed, studied or otherwise looked at from afar.

So a magazine that publishes stories about island people does so with a certain degree of risk. The danger isn't so much that people will take offense; an editor with reasonably sharp eyes can keep the libel, slander or otherwise unflattering characterizations out of the pages. No, the risk is more subtle: if you persist in "describing" islanders in the terms of an anthropologist — looking at them, so to speak, from afar — you have a very good chance of not being taken seriously. In this business, that's worse than being sued for libel, which can merely cost you money.

Of the nine stories included here, one exhibits more "edge" than the others. The late George Putz, after a mere 20 years on and around Vinalhaven, undertakes a thorough analysis of how islanders think. He succeeds brilliantly, becoming the exception that proves the rule.

The other stories here are less daring. Several are accounts of islanders' own experiences — island life from within; others explore island life in journalistic fashion; still others through memory. Each of these approaches is safer, perhaps, than the out-there characterizations attempted by Putz, but they're no less valid.

David D. Platt

The Islander (detail), Jamie Wyeth, 1975
Above: Peter Ralston

On Islanders

George Putz

Inhabited islands are something like aquaria. They are capable of self-maintenance, but like an aquarium, their functional components lack diversity, and are continually stressed. Compared to a continental community, there are fewer eco-options or strategies by which an island community can adjust to changes. The kinds of history Maine islands are experiencing today are not evolutionary. They are impulsive and vigorous, and for all the difficulties of adjustment people are having in mainland towns, they are nothing compared to those being faced by islanders. That an island's human community is like an aquarium is not simply a metaphor or analogue, for social life on an island is quite literally life in a gold-fish bowl.

The list of islander characteristics is familiar; but in each case the features are written larger and more stark in an island context:

Independence — small boats and social circles demand it if a personality is to survive.

Loyalty — ultimate mutual care and generosity, even between ostensible enemies.

A strong sense of honor, easily betrayed.

Polydextrous and multifaceted competence, or what islanders call handiness.

A belligerent sense of competition, interlaced with vigilant cooperation.

Traditional frugality with bursts of spectacular exception.

Earthy common sense.

Opinionated machismo in both the male and female mode.

Live-and-let-live tolerance of eccentricity.

Fragile discretion within a welter of gossip.

Highly individualized blends of spirituality and superstition.

Above: Peter Ralston
Skeet MacDonald, Isle au Haut

JEFF DWORSKY

A complex oral tradition, with long memories fueled by a mix of responsible record-keeping and nostalgia.

And finally, a canny literacy and intelligence.

Until the late 1950s the maritime economic focus of the island I moved to fit the traditional values and character most people associate with self-reliant islanders. Central to this fitness was the ability to create and maintain virtually everything required for an equitable and ordinary life — houses, food, clothing, boats, equipment and gear, entertainment and fun. If you could not make it yourself, it was readily acquirable by trade in time and kind.

Many things conspired to end this fit; most particularly advances in several mainland technologies. Small gas engines grew into the Chevy-Olds family of V-8 engine blocks; there was an explosion in electronic navigation and communication technology; and, finally, synthetic fibers, coatings, and plastics all helped encourage an irreversible shift from labor- to capital-intensive efforts. And with capital comes accounting, and a vastly more complicated participation in the world of the mainland. What you once made, you must now buy. What once you repaired, is now placed in the hands of a specialist.

Islanders are also adapting to rapid changes in the seasonal community, and the inevitable growth of tourism.

Caretaking and maintenance jobs become increasingly available options, though attended of course by a general propensity to care less about the property of others, as increasingly "others" exist outside the circle of community sentiment.

I suspect most readers of this publication love islands and have a deep place in their heart for their inhabitants. Yet many of those same hearts believe that conventional seasonal tourist-based development can benefit Mane island towns. Yet, for the life of me, I can't see how that can work. Tourism, in particular, is a direct anathema to everything that allows islands to function with community integrity. It's one thing to live in a goldfish bowl with neighbors that share generations of curious history; it's quite another to be behind an aquarium glass for the entertainment of total strangers.

My own suspicion is that the growth of seasonal tourism fundamentally betrays a sense of pace; of island time.

Islanders feel isolated, because they are isolated; and the consequent reticence and sense of irony that comes out of this isolation was much more compatible with the aspirations of the rusticators and pilgrims of old, than with the active, can-do, do-good inclinations of the modern visitor, seasonal resident, bored-but-hyperactive retiree, not to mention full time midlife transplants with firm notions of what's good for other people.

Hospitality has always been nearly a religion on islands, but it now must compromise itself and become ever more choosy as islanders have to know in advance, for instance, which visitors appreciate the smell and sounds of maritime work.

It is, at the very least, bad manners to care about any aspect of an island and its phenomena,

JEFF DWORSKY

without caring also about its people — past, present, and future; not with sympathy and patronization, but rather with empathy and plain honesty. It is for good reason that many, if not most, conservation and preservation efforts are viewed as a kind of class predation by islanders. Until interested personnel in conservation and research understand islanders, whole realms of knowledge and experience simply will not be open to study. Ninety percent of the useful information to come out of island study is known already, scattered through the hearts and minds, attics and backyards of islanders. Much of the vast material is yet to be discovered by researchers.

What is at stake in the midst of the furious pace of these changes?

First, islanders are among the last Americans who as a group can presume their sense of place;

Second, islanders are hunters and gatherers by tradition and instinct. Within bounds of strictly understood codes of decency, the world and its inhabitants are resources.

Third, and in apparent contradiction to the above, islanders are extraordinarily good and assiduous nurturers and husbanders. The usual outsider's impression that commercial fishermen are strictly exploiters fails to notice the fact that lobstering has become mariculture.

Fourth, the feelings of isolation and the uncanny sensitivity to signs of a change in the weather and what it can bring, gives islanders an unusually keen sensibility to natural history.

Fifth, there is a self-consciousness about islandness among islanders. I call it cellarhole melancholy. This is a generalized sense of loss, of what could have been, of what probably happened that shouldn't have; of the blood, sweat, and tears that permeates every foot of island rock, soil, and beach.

Sixth, island institutions are deep and traditionally effective, for they seldom operate solely for the advantage of their members. What is gained in them is a celebration of identity and fellowship. They are a blend of romanticism, of oral literature, of forum, of unity in rites, of security, sharing, of wit, art, commiseration — all the truly important things in life.

Seventh, there are the heritages, used symbolically on islands on a daily basis with all their myths, habits and stories, incumbent skills, traditions and uses. There is a lifetime of study and work in these heritages: fishing, boats, and boatbuilding. Quarrying and stonework. Farming and forestry. Architecture and community design. Arts and literature. Trail and shores. Hunting, gathering and folk crafts. Even science. For all island towns have their archaeologists, rockhounds, birders, flower pressers, woodlot managers, whale watchers, and so on, and many of them are doing first-class work. Islanders share a common sense with other islanders worldwide, in much the same way that scientists share a common sense worldwide on such things as denotation, logic, control, and proof.

America needs her islanders. It is, of course, fatuous and arrogant to speak of preserving island communities. The die is cast and history shall have its way. But there is still in the Maine archipelago, and on islands elsewhere, an intact vision of the world which differs from that of others and which offers not merely diversity and its advantages, but a sensibility about the world that the world could use, since citizens everywhere are coming to realize that the earth itself is an island. In this sense, mainlanders are the pre-Copernicans, and islanders are the most sophisticated, modern, and up-to-date. Islanders know about islandness and all of us should have some of this imprinted on our consciousness.

1984

The Coot Hunter (Detail), Andrew Wyeth, 1941

COURTESY OF THE ART INSTITUTE OF CHICAGO

The Legacy of Cyrus Rackliff

Amy Payson

Ever since I was a little girl, I remember hearing stories about my family, but the story of my grandfather, Cyrus Rackliff, and his accident on Green Island was, to me, typical of life on the Maine coast at the time. Although I don't remember my grandmother, Emily, who was Cyrus's wife, my mother and father (Archie and Edna Rackliff) lived with Emily at Cyrus's homestead all of their married life. My mother and Emily shared their thoughts and feelings while they did the mountains of washing and the cooking for the family.

COURTESY OF AMY PAYSON

Emily described her early life on the islands, especially Metinic Island, where she had been a teenager. There she met Cyrus, who was living on Green Island (Metinic was within rowing distance of Green, also the closest piece of land). Emily Foster was a striking, dark-haired beauty, a prize for any man, and Cyrus was "the boy next door." When they married in 1861, Emily moved to Green Island and lived with Cyrus and his mother, Lydia. Then Cyrus and Emily, with two children of their own, moved to Dix Island in the Mussel Ridge, where they ran a boarding home for the stonecutters working the quarry.

Because the story of Cyrus's accident meant a lot to me, I asked my mother to write it down. I kept this handwritten account safe with my choice possessions. One day I went to read the story and couldn't find it. Embarrassed that I might have lost the treasured story, I finally asked my mother to write it down again, but I didn't keep after her to see if she had done it. So, when she died I badly regretted all that was lost, including Cyrus's story.

One day I was going through her things and came across a book, *Lighthouses of the Maine Coast*, which my oldest son, David, had bought for her. She told me that she wanted David to have this book back when she was "finished" with it. When I opened the book out dropped a folded copy of "Cyrus's story," handwritten in rough draft form that my mother had done following my request!

This is the story my mother told of Cyrus's accident:

One particular day Cyrus and the boys were going seabird shooting on the ledges just outside Big Green Island. They wanted to be there to set their tollers before the birds came in about daylight to feed. In the morning when he woke, Cyrus told Emily that he didn't think he would go sea ducking, since he had had a dream during the night that he was breathing shot. He had awakened choking. But a little later, it turned into such a nice morning that Cyrus decided he would go shooting anyway. As he got his leather boots out to grease them, an old clock on the shelf that had been silent for years struck. Cyrus's mother was very superstitious, and, fearing something terrible would happen, begged him not to go. But Cyrus was bound and determined to head out for the ledges.

He put his gun in the dory, but as he got in, the gun, which for some unknown reason was cocked, went off. The heavy load of birdshot struck him below the knee. It was a bad wound, so Cyrus's relatives sailed him across Two Bush Channel and up the Mussel Ridge to Ash Point, where an uncle took him to a doctor in Rockland.

Cyrus Rackliff

Cyrus's mother, Lydia, viewed the accident as an ill omen and told Emily, who was pregnant at the time with her first child, that the baby would probably be "marked" because Emily had to dress her husband's gunshot wound. Imagine the worry to a young bride and the relief that her son, Elmer, was born without a blemish. Cyrus kept on farming and hunting even with his bad leg, which Emily washed and dressed daily. Sometimes, when she changed the dressing, scraps of his old pants leg would appear, along with an occasional piece of lead shot.

After 20 years of suffering, Cyrus finally decided to have his leg amputated below the knee. Had he not had such a strong constitution, he wouldn't have survived. With a bottle of whiskey to kill his pain, the doctor took off the leg in the front parlor of the house where they lived. After this rough operation, Cyrus asked for a chopping block and performed an autopsy on the foot to see what had caused him such agony for those 20 years. I guess he expected to find sharp bone fragments that would account for the pain, but nothing was found to satisfy his curiosity.

1986

PETER RALSTON

ISLANDERS

Winters when we set our traps offshore,
we saw an island further out than ours,
miraged in midday haze, but lifting clear
at dawn, or late flat light, in cliffs that might
have been sheer ice. It seemed, then, so near,

that each man, turning home with his slim catch,
made promises beyond the limits of his gear
and boat. But mornings we cast off to watch
the memory blur as we attempted it,
and set and hauled on ledges we could fetch

and still come home. Summers, when we washed
inshore again, not one of us would say
the island's name, though none at anchor sloshed
the gurry from his deck without one eye
on that magnetic course the ospreys fished.

Winters, then, we knew which way to steer
beyond marked charts, and saw the island, as
first islanders saw it: who watched it blur
at noon, yet harbored knowing it was real;
and fished, like us, offshore, as it were.

Philip Booth, 1986

TERRY BARNUM

Bob and Helene Quinn

"It's the Ancestors"
The Quinns of Eagle Island

Todd Cheney

Eagle Island lies in a string of islands fringing the deep waters of East Penobscot Bay. On the island's northeast corner, the confluence of tide and current drive a timeless surge against 60-foot cliffs that rise black and sheer from the rockweed and water. At the top, the igneous rock wall curls back under a mat of soil crammed with the roots of stunted trees, their branches lopsided, growing away from the prevailing wind. Follow the cliffs to the west, and they subside to a beach that intersects the water at a gentle angle, in a sweeping crescent smooth as a curve of the moon. Spruce forests flank the beaches and cliffs everywhere, and the landscape is wild and primitive….

On Eagle for a long weekend, I walk up the hill on an early autumn evening to Quinn House and find, from the vantage of the high field, a show of sublime and contrasting lights. To the east, the sea, land and sky are fused in shades of blue and black; to the west, the water still shimmers with orange reflections of the sun, half-sunk in the Camden Hills and throwing shadows of the spruce trees across half the field. Imperceptibly these shadows thicken. The field and trees blacken as the night spreads over the great curve of the land, and the ancient light of the stars begins to brighten the sky. I find Bob in the kitchen shadows, peeling potatoes; while waiting for them to boil, he pours a bowl of Grapenuts and smothers them with coils of dark molasses. "I'm not one to cook much," he says. "Get so hungry I can't wait for things to get done."

Robert Louis Quinn was born July 13, 1939, and so lived on Eagle only two years before his family

PETER RALSTON (2)

The Eagle Island school, abandoned in 1941.

moved to Camden. It was, finally, the problem of schooling that forced the fourth generation of Quinns to remove to the mainland. It was the practice for Eagle and other islands to send high school-age children to board with relatives or friends during the school year, but Erland Quinn, Robert's father, didn't like the idea of sending his children away to live with another family, and he made the decision in December, 1941, to take his family to Camden. With the closing of the school, the few remaining families had no choice but to follow the Quinns ashore.

Out of all the family Bob Quinn is the most intimately attached to the island and says, "I never had any doubt, from day one, that I would do anything else." And Helene, his wife, says, "For Bob, there isn't any choice." He spent his childhood sum-mers on the island, staying with his grandmother Hattie, and after high school he moved there and made a few dollars trapping mink.

Helene grew up in Warren, Maine, where her grandfather, Charles Howard, had moved the family farm from Eagle, and where her father, Richard,

had been a successful dairy farmer. She first came to Eagle when she was eight to visit her Aunt Marion at the Howard Place, but by the next time Helene came to Eagle, Marion's house had burned and from then on families would stay "mixed up together" at Quinn House after Bob and Helene married in 1966....

Bob's first 30 years were a time of deep decline on Eagle. There was no economy, and, except for a cow and some chickens at Quinn House, even the farming had ceased. A 1972 photograph of the old homestead-hotel shows it sagging and gray. It hadn't seen a coat of paint in 40 years, the roofs leaked, windows were falling out, sills were rotting, the boathouse and wharf threatening to fall into the bay, and each year the fields shrunk and the spruce grew higher, shutting off more of the island world to which Eagle belongs.

A visitor to Eagle once characterized Bob as having "the hands of a fisherman and the mind of a poet," a mix of practical and visionary inclinations joined together with the ability to direct thought into actions. I once asked Helene how the family managed to stay on the island when all the families on neighboring islands had disappeared. "It's the people," she said, "they hang on when there's no reason to hang on...."

There's little doubt that if not for Bob Quinn the whole family presence on the island would be gone by this time. His energy has rallied the extended family's interest and participation; many who once turned from the decay and isolation have turned back and now see the island and Quinn House as something solid and unchanging in their lives. Quinn House has received life-saving structural repairs, interior restoration and a new coat of white paint on the old clapboards. The barn has been roofed, and the march of spruce and juniper into the fields has been turned back.

The demographics of the Maine coast for more than a hundred years have followed a pattern in which the natives sell their land to people who live where there's big business and big money. Set against that erosion of tradition is Bob's goal of keeping intact the property, the customs and the history so they will be here for generations to come. He spends a lot of time fretting this mission, and following through on means to carry it out.

For now, however, I can say that the Quinns are more thriving than threatened. There is yet the feeling that something wise and more powerful is close by, something warm and human and comforting of a sort the rest of the world holds in short supply. It has to do with the past, of being conscious of its presence, of its meaning for our lives. It's something to do with what Helene feels about the island when she says, "It's the ancestors. We talk like they're still here. Our being here keeps them alive."

1987

Eagle Island, looking north

Avelinda

The Legacy of a Yankee Yachtsman

Tom Cabot

Installments of Tom Cabot's memoirs of four decades of cruising the coast and islands of Maine were published in volumes 7 and 8 of Island Journal. *Later, in 1993, the Institute published Cabot's complete account of his Maine cruising memoirs in a book titled* Avelinda, *after the name of his favorite vessel.*

By the spring of 1910, when I was about to have my 13th birthday, one of my uncles evidently thought I was old enough to have a sailboat of my own. He persuaded my grandmother to give me a Manchester one-design knockabout as a birthday present. It was the happiest day of my young life…. To be sure, my yacht was only 17 feet on the waterline, but it had a self-bailing cockpit and a cuddy with two wooden bunks. I named the vessel TULIP and put out a mooring for it off West Beach in Beverly Farms where we were spending the summer. In early July, during a northeast storm, the screw shackle connecting the mooring chain to the pennant let go, and, to my dismay, the vessel came in on the beach where the surf banged it to the point of breaking two ribs and some of the planking. My mother put up the money to salvage the vessel and have it repaired. It was a lesson I never forgot. In a lifetime of sailing I have, of course, had other vessels grounded, but this is the nearest I ever came to a complete shipwreck…

As summer approached in 1919, I found myself with a few weeks free before summer school started at Harvard (I was hurrying to complete my degree so that I could get married), and with my college roommate, Alec Bright, I cruised eastward in TULIP from Beverly Farms as far as Frenchman's Bay in midcoast Maine. It was a memorable trip. I polished my skills at piloting and celestial navigation and learned a lot about the perils of anchoring and how to avoid dragging the anchor if the wind freshened. We also had lots of experience finding the way in dense fog.

To prepare TULIP for the trip, we had a large wooden box nailed down to the cabin sole in which we stored our provisions, cooking utensils, a small Primus stove, and our spare clothing. We had a kapok mattress which roughly fitted the cockpit, a sailcloth awning to stretch over the furled sail at night and tie down to small cleats on the gunwale at either side, and another piece of sailcloth and a couple of blankets to sleep under. All of these were stored below decks during the day.

With a fresh breeze at our backs, we set off from Beverly Farms late one afternoon and rounded Thatcher Island off Cape Ann before dark. We passed to seaward of the Isles of Shoals and Boon Island

PETER RALSTON

COURTESY OF CABOT FAMILY

Tom Cabot on AVELINDA

with its tall lighthouse. Shortly after midnight, the wind being aft, we were carrying the spinnaker and mainsail. While I was napping, with Alec at the tiller, he suddenly let out a yell and put the helm hard down. He thought he had seen breakers ahead. The spinnaker came aback and we had some difficulty getting it down. While we wallowed in the sea with the mainsail flipping, we peered ahead, but I could see no sign of breakers so we resumed our course toward the northeast.

Between 1:00 and 2:00 a.m. it began to rain. In pitch dark we put on our oilskins. The wind began to slacken, and by dawn it was a flat calm. Our vessel was rocking severely in the remaining waves and we sat there, cold and mildly seasick, eating only a few bites of cold biscuit for breakfast. By noon the seas had subsided. We were still miles offshore and there was still no wind. We decided to try to tow our boat toward the land with the dinghy. With one of us at the helm and one rowing, we took turns at towing shoreward. We could see no recognizable landmarks on the shore and didn't know where we were. It was nearly dark before we got close enough to shore to identify some lobsterboats moored in what looked like protected water, and it was quite dark by the time we got among them and were able to anchor. We had only a small kerosene ship's lantern with Fresnel prisms, which gave too little light for us to find anything much to eat or to bother with cooking. Having been awake for 36 hours, we had no trouble sleeping.

The next morning, fishermen told us we were in Potts Harbor (near South Harpswell). It was a bright day with a good breeze and we sailed eastward around Cape Small, inside of Seguin Island, and came into Port Clyde in the late afternoon in plenty of time to cook some canned stew for dinner and have a walk ashore before dark. The next day we sailed on through the Muscle Ridge Channel to North Haven for the night, and the following day through the Deer Isle Thoroughfare to Burnt Coat Harbor on Swan's Island. We went ashore for some fresh milk and bread at the small store on the shore near the wharf, which is now the fishermen's cooperative. The two following nights were spent in Northeast Harbor with a full day's sailing among the Porcupine Islands in Frenchman's Bay. We returned by way of Eggemoggin Reach, where we spent a night anchored off the north shore of Deer Isle, not far from where the large suspension bridge now serves that island.

There was thick fog the following day. We missed a buoy and got lost. We found ourselves among ledges and hit one lightly with no damage. We tacked into a light southwesterly wind all day in the fog, not knowing where we were but occasionally seeing an island shore, and finally anchored in the lee of a wooded island for what proved to be a rather restless night. It began to rain shortly after dark and by midnight the wind was freshening into a storm. We had only about 15 fathoms of half-inch hemp rope for our anchor, which was the old fisherman's type. None of the modern patented anchors had been invented. The depth of water was much greater than we had anticipated, and although we were close to the shore, there was very little more than enough rope to reach bottom. About midnight we realized that our anchor had dragged and we were adrift in deep water.

The longest rope we had was the peak halyard so we unrove it, attached it to the anchor rope, and got the vessel head to wind again, but we couldn't be sure in the dark whether or not we were still dragging.

By dawn it was still raining but the fog had cleared. After some study of the chart, we found our position to be between Great Spruce Head Island on the west and the Barred Islands on the east. There was a large house on the northeast corner of Great Spruce Head Island, and we decided to row to it.

All our clothes were soaking wet and we were miserable and cold. We wrung what water we could out of our wet underwear, put on oilskins with nothing but underwear beneath, and in short

$\mathcal{A}$velinda

The Legacy of a Yankee Yachtsman

Text and Photographs by
THOMAS D. CABOT

PETER RALSTON

Tom Cabot on Butter Island

order made it to shore. It was about quarter of seven in the morning, and a young boy and a girl our age were playing ping-pong on the screened porch. They asked us in, lit a fire in the living room, and invited us to stay for breakfast. It was the Porter family from Chicago. It was Nancy, the oldest, who had been playing ping-pong against Eliot, her brother. Two younger brothers, Fairfield and John, soon appeared with the parents. We were much embarrassed, having only underwear under our oilskins, so before breakfast we rowed out to our vessel, got some more wet clothes on, and rowed back for a meal. We wound up staying all day and spent the next night in the shelter of their harbor, leaving the following morning to sail to Tenants Harbor.

That was our first cruise of Maine. I can't possibly remember the scores of cruises we had later and all the places we anchored, but I can clearly remember some of our misadventures and many of our favorite harbors and gunkholes....

Our first family cruise in a larger vessel along the Maine coast was in the summer of 1931 when we chartered the schooner PORQUE NO, a vessel out of Camden. On the first day we sailed over to Great Spruce Head Island and anchored in the private harbor of the Porter family. Before dark, John Porter came alongside and told us that there was a radio prediction of high winds before midnight and that he thought we would be less exposed if we anchored in Barred Island Harbor nearby. He offered to pilot us there. It was low tide and twilight. On entering the harbor we hit on a sunken ledge halfway between the northernmost island of

that archipelago (then called by the fishermen Peak Island but now called by the family Escargot) and Western Barred Island. We were soon off the ledge and anchored in the harbor of the night. It was our first night on the vessel and there were only four berths below deck. There being five of us, Tom Jr. was nominated to sleep on deck. He had a mummy-shaped sleeping bag with no zipper. He was only eight and not a strong swimmer. About 2:00 a.m. I was awakened by a call. I thought he had called in his sleep but a moment later I heard splashing. I rushed on deck. The rising tide was streaming by the vessel, and in the wake I could see astern something on the water. I dove for it; when I came up, I had only an empty sleeping bag. In a panic I started yelling hysterically. While the rest of the family swarmed on deck, I splashed around trying to find my son. After what seemed like hours, someone heard a faint cry from the bow of the vessel and there was Tom Jr. hanging onto the bobstay, the only part of the vessel that he could get a hold of from the water. He and I were both pulled on deck with his wet sleeping bag, and he was put in my warm bed below deck while I was relegated to sit with dry clothes on the deck for the rest of the night. The afterthought of that near drowning haunted me. From terror or cold, I'm not sure which, I shivered through the remainder of the night. In the beautiful dawn, I was near weeping with emotion. It seemed the most beautiful dawn I had ever witnessed and I resolved then and there to try to buy the surrounding islets.

1990

When You Live on an Island

Dean Lunt

There was no great revelation to move off of Frenchboro: I just went to college and went after a job. But I loved it there. I still do. I never felt the suffocation or the desperation to get off the way some kids from islands felt. I'd love to go back there if I could make a living. But even now, I haven't accomplished all I want to do....

You gain or inherit a sense of community when you live on an island. A lot of towns don't have that anymore. You gain an identity from an island. Each one's different. But you always have that unique identity to draw on. There's also a sense of security on an island — very few outside forces affect the community. Frenchboro has one harbor and most of the houses ring the harbor. If a boat comes into the harbor, you know it: you know who's coming and going. It's safe. On the island, I never had any boundaries from the time I could walk....

1997

PETER RALSTON(2)

Margaret Wise Brown
"A Writer of Songs & Nonsense"

James Rockefeller

Islands can be personal castles or prisons depending on how one views their moat of water. To Margaret Wise Brown, her place on Vinalhaven was a castle of fairy-story proportions. Margaret brought me there to the head of Hurricane Sound in the summer of '52. She called it The Only House because looking out at night, more often than not, no other light was visible.

No road existed. The surrounding forest was yet another barrier against the outside world. Our entry was the little house of Mildred Brewster and Maynard Swett in that small drain behind Strawson's Point. Mildred had done the cooking in the Boarding House at Wharf's Quarry for the men back when, and Maynard, over 70 years old, lobstered out of his white peapod with the green gunwale. Here Margaret kept a gray flat-bottomed punt built by Skoog of Carver's Harbor. It comfortably held the two of us, her Kerry Blue terrier called Krispin's Krispian, groceries, a case of wine, and other household necessities. The 20-minute pull up the Sound was a pleasant interval on that warm and sparkling day of my arrival, gently pushed along by the southwest breeze.

Dog and Margaret occupied the sternsheets. Krispin glared at me while I eyed his mistress. Krispin was disagreeable by nature, but then, in all fairness, it was not easy for him being in the proximity of another male who also loved his mistress. Margaret wore her usual working costume of white slacks, espadrilles, and a blue blouse open at the neck. Her straw-colored hair, tumbled by the breeze, was a perfect frame for those crinkly blue eyes that looked at you, with you, through you, while absorbing everything within 360 degrees. She trailed one hand in the water, lifting it eventually to extend a dripping fin-

Margaret Wise Brown with Krispin's Krispian on Vinalhaven

JAMES ROCKEFELLER (3)

ger to a passing dragonfly. To my amazement the insect landed as if it had no choice.

"Warlock," she said to me, "what must it be thinking, flying over all this bright blue water? Must be the lobster buoys are a flower garden?"

With Margaret you lived an ongoing series of mini adventures. Involved with the smallest event, she pulled her companion along into a magical world she composed on the spot. She called me her Warlock because I wore a beard back then and could look very fierce when being protective.

Too soon we arrived at a tiny beach hidden behind a long chunk of rounded granite. Entrusted with the case of wine, I walked up through the long grass of the tiny meadow dappled with hawkweed toward the tiny house. The high-pitched roof, black attic window, black-framed windows of the lower two floors, the gray weathered clapboards, made it both intensely appealing yet mysterious. It was as diminutive as a child's playhouse, but one sensed immediately the inhabitant was neither a child nor a casual rusticator.

Access was gained by saluting an ancient pear tree and mounting steep steps, almost a ladder, that teetered upwards to a circular porch 15 feet above the ground. This platform, in turn, was guarded by a granite ledge to the west that resembled a smiling whale, and an apple tree that intruded over the railing to the north. Ice cream parlor chairs and table formed an eyrie for eating, talking, or just surveying the warblers, woodpeckers, ospreys,

gulls, and terns who considered the place their own, which delighted Margaret. The long, narrow, steep steps were yet another psychological barrier against those things and beings beyond the forest and the bay. As Margaret put it, "Here I am far away from the fidget wheels of time," talking about her frenetic winters spent in New York with agents, publisher, and her host of social commitments.

From this roofless treehouse you entered a tiny kitchen, off of which opened an eight-by-ten room that held a love seat, a reclining couch, potbellied stove, and a long table in front of the window where Margaret did her writing. To the left of her writing station was a door that swung out onto nothing but a ten-foot drop to the ground below. It bore a brass plate saying Belle McCann. Belle was a previous owner before the house had been raised and a floor added underneath. Off the sitting room was an even smaller bedroom with a brass bed and dresser. The whole place was the size of a ship's cabin.

The window was an inspirational place commanding Hurricane Sound with its myriad little spruce-tipped islands. This view inspired *The Little Island*, perhaps her best-known book.

Kerosene lamps were the sole illumination. A rose-colored globe hung over the tiny kitchen table on an adjustable chain. Another glass lamp, this one ruby-tinted, lit the writing table, while two companion pieces moved about as needed. A pair of exquisite small Italian rococo candelabra created a flower display on the vertical paneling between the sitting room and bedroom, adding more soft ambiance when darkness fell.

Of an evening, with perhaps a red spaghetti sauce laced with garlic bubbling on the kerosene stove, the red wine in goblets, Margaret would seat herself at the table that had witnessed many things, her eyes shining in the fairy story light, and was definitely the queen of this special kingdom. I say "queen," for everything in the tiny house appeared her personal subject, chosen for shape, for color, for adding catalytic quality to the overall sense of a cozy den, yet in such an unstudied way as if to be a natural extension of herself. Eggs were stored in a bowl to enjoy their shape and facilitate their use. Wildflowers winked from glasses, cups, vases, or copper pots. The sublime imbued the whimsical with a dignity that I likened to their originator.

Margaret loved fur. "Remember, we are animals," she was wont to say. Rabbits were a special totem. She had long eyelashes and would often accent her eyes to give herself an almond bunny look when feeling mischievous. Many friends called her "The Bunny," and she often referred to herself as "The Bunny No Good"

Margaret Wise Brown's 'Little House' on Vinalhaven

when up to some lark or saying things like "I'm going to give all the birdbrains egg cosies for Christmas."

There was a lot of fur around: a fur rug on the floor, fur on one of the couches, a fake leopard skin covering on the bed. She was very proud of the fact that the English Queen Mother reputedly kept *The Little Fur Book* (it was covered in rabbit fur) on her bedside table.

After a few days with "The Bunny," you weren't sure whether people acted like animals or animals like people. As one's eyes are drawn to those of a wild animal to gauge their intent, so mine would often gravitate to hers. There was always more going on in there than the viewer could ever grasp. The look would vary from youth to venerable age to childlike wonder, mischievousness, gaiety, somberness, or the wisdom of a seer.

"No one will ever know my age," she said laughingly one day. "How could they? It keeps changing."

In earlier years — before her time — goats, chickens and a cow lived downstairs. Now there was a workshop and guest room. The latter also served as a gallery for her paintings. She explained that from early on she knew she could either be a credible painter or a writer and had decided on the latter, so painting became a hobby. One oil was of a white dog (with rabbit-long ears) lying on the love seat upstairs. Through the window in the

painting peeped The Little Island. Another featured the horse weathervane she had whimsically mounted on the end of the stone wharf; yet another showed a white china water pitcher filled with flowers. The last conveyed a different mood. The Only House stood somber in its black trim under a lowering sky. A small drab figure huddled against the stoop. This she had done after the death of her dear friend, the poetess Michael Strange, wife of Barrymore the actor.

"When I can no longer write, paint, or read, that is the end," she once told me...

Walking up the granite escarpment behind the cabin, one came to a flat circular expanse of stone some 60 feet in diameter. As we approached, she would press fingers to lips for silence, for here was the Fairy Ballroom where the "little people" danced at midnight overlooking Hurricane Sound. It was always just possible, even in broad daylight, there might be one peeping from behind a bayberry bush.

I loved to go up there of an evening, for we would stand on a rock outcropping and watch dusk enfold the Sound. Dawn and dusk were important times for The Bunny, as was the languor of noon, the rising of the moon, storms, and calms. Standing on the promontory, outlined against the darkening bay, she radiated the elemental dignity of a wild animal free in its native habitat. Often those eyes of hers would go far away where no one could ever reach, and one evening she turned suddenly and said, "We are born alone. We go through life alone. And we go out alone." I never have forgotten that moment, painful as were the words, for what she said was true. She saw herself in a frame where human beings were but one component of a larger tapestry.

In the woods and fields Margaret moved like a deer. She told of going beagling and running with the hounds for hours on end. During berrying excursions she could wriggle through the most impossible of tangles at incredible speed and eat berries off the bush like a bear I once had. A herring fisherman who set his nets out front once said to me, "That Margaret! If you saw her in the woods come November and she was wearing horns it would take a steady mind not to shoot." Then he added with a wistful grin, "I'd rather take her home alive, myself."

For venturing on The Bay, Margaret had a treacherous North Haven Dinghy. One day we had a wonderful sail down to Hurricane, trailing the bottle of white wine behind on a string with Margaret puffing on her pipe, reciting one of her many lyrics, "The Fish with the Deep Sea Smile." The ballad begins:

They fished and they fished
Way down in the sea down
*   in the sea a mile.*
They fished among all the fish
*   in the sea*
For the fish with the deep sea smile.

On the way home the southwest breeze turned into a small gale. We rushed along faster and faster until at the end of Leadbetter Island the dinghy sailed her bow right under. There we were with the sail up, going nowhere, paddling around in the cockpit. I was mortified, considering myself something of a sailor. But Margaret puffed away on her still lit pipe, asked if there was any wine left in the bottle, and giggled with glee. Just then Goldie McDonald, the guardian of Dogfish Island, happened by and pulled us, dripping, into his boat. Goldie was one of Margaret's favorites. She even used him as a pen name.

Goldie took one look at The Bunny with her wet clothes clinging to her athletic frame and said with feeling, "Gawd, Margaret, you look better wet than dry!"

She laughed all the way home. In her eyes it couldn't have been a more perfect day.

Aside from sitting in the evening bathed in that ruby light, going up to Wharf's Quarry was my fondest memory. Carrying a hamper, towels, and soap we would take the path around the back cove, plough through an overgrown meadow and

tangle of brambles, then walk under a canopy of huge spruces until coming to a granite ledge lying in the gloom like a forgotten Stonehenge. Over this we pushed and pulled ourselves, emerging on a gently sloping expanse of stone which we followed upward until standing atop the quarry. There we would gaze down 50 feet of sheer rock wall to the pool of water with the pyramid of grout on the far side. To the left stretched Hurricane Sound and straight ahead to the west was Leadbetter's Narrows with a backdrop of the Camden Hills. The entry to our destination was at the far end where the granite sloped down to cattails, with stone and vegetation arranged as if by an artist's brush to form a hidden water garden. Here on a flat rock by the water's edge we would spread our things and have our biweekly ablutions. Afterwards, drying on the sun-warmed granite, we would eat lunch in almost mystical serenity.

Margaret would talk about her writing, and I, our future life together. She had published 72 books to date, "with nothing serious to say," as she put it. Little did I comprehend at the time what a pioneer she had become, and how revered in the writing of children's books.

"Warlock," she would muse, "someday I would like to write something serious when I have something to say. But I am stuck in my childhood. That raises the devil when one wishes to move on."

"What do you want to put on your tombstone then, if not recognition for children's works?" I said facetiously, little knowing that only a few months hence she would lie dead in France of a blood clot.

She thought a bit, watching the white clouds pass overhead before turning and saying in all seriousness, "You will put 'A writer of songs and nonsense….'"

1994

JEFF DWORSKY

A FAREWELL

For a while I shall still be leaving,
Looking back at you as you slip away
Into the magic islands of the mind.
But for a while now all alive, believing
That in a single poignant hour
We did say all that we could ever say
In a great flowing out of radiant power.
It was like seeing and then going blind.

After a while we shall be cut in two
Between real islands where you live
And a far shore where I'll no longer keep
The haunting image of your eyes, and you,
As pupils widen, widen to deep black
And I am able neither to love or grieve
Between fulfillment and heartbreak.
The time will come when I can go to sleep.

But for a while still, centered at last,
Contemplate a brief amazing union,
Then watch you leave and then let you go.
I must not go back to the murderous past
Nor force a passage through to some safe landing,
But float upon this moment of communion
Entrance, astonished by pure understanding—
Passionate love dissolved like summer snow.

May Sarton, 1992

KOSTI RUOHOMAA

Brothers on
the Rock

Colin Woodard

In 1930, at the start of the Great
Depression, Doug and Harry Odom got
their first look at Monhegan Island.
Their older brother Sidney, 22, had been
dispatched by his boss to run the island's lit-
tle summer store and he took Doug and
Harry, 16 and 14, up from Quincy,
Massachusetts to help him.

"I took one look at it and said, 'What the
hell am I doing on a rock like this?' " Harry
recalls. "I told my brother I wasn't impressed
at all. But when the fall came, I hated going
in! As a kid, to take trips and walk all over
that island. Oh, it was a great place." The
boys couldn't wait to come back.

When the Odoms first came out, there
were four Model T Ford trucks and three
crank telephones on the island. Their store
had one of each, along with a small Delco
32-volt generator and a bank of batteries that
powered the lights and a tiny freezer.
Everything else in the store was refrigerated
with blocks of ice cut from the ice pond up
on the hill behind the lighthouse.

Throughout the 1930s, the store
remained a seasonal operation. Come winter,
Harry went inshore to work as a salesman
for the Rockland office of the John Bird grocery chain, while Doug was employed by the
Kennedy Butter and Egg Stores in
Massachusetts.

As the clouds of war gathered, Doug
took full-time work operating cranes at the

The Odom brothers in their Monhegan store

COURTESY OF BEN ODOM

naval shipyard back in Quincy, while Harry tended the store. But within two years the brothers were 12,000 miles apart, fighting in battles on opposite ends of the planet.

Doug served in the engine room of the USS MASSACHUSETTS, a 35,000-ton battleship he'd watched being built while he was perched atop a shipyard crane. The MASSACHUSETTS was a fortunate ship. Doug recalls there being not a single combat death among the ship's crew of 2,500, despite taking two shell hits during a naval battle outside German-occupied Casablanca and three years spent in the Central Pacific providing support to amphibious landings against the Japanese. Doug was aboard the MASSACHUSETTS for more than three years, returning to Quincy after the Japanese surrender.

Harry was drafted into the Army, a staff sergeant in a unit that engaged in heavy fighting in Sicily and northern France. He was wounded by shrapnel shortly after the Normandy landings and spent two weeks in an English hospital; he returned to find that many of his buddies had been killed. Shortly thereafter, while occupying a knoll on the front, Germans overran his position. Harry was captured and shipped to a prisoner of war camp in German-occupied Poland where he spent the next six or seven months surviving on meager rations: a bowl of potato soup and a piece of bread once a day.

"They didn't abuse me but it was really rough not having enough to eat," he recalls. "I used to be a chowhound and boy, did that hit me hard."

"I'd been sleeping on the ground, in box cars, on hard floors," recalls Harry, who received four medals including a Purple Heart. "When I got back to my mother's she said, 'Oh, you sleep in here,' a room with a bed with all those soft, thick mattresses. I couldn't sleep in it. I took a pillow and laid on the floor."

But while Doug and Harry were off fighting the war, the store had gone into debt. "They thought they'd run the store and get out of debt and then go their separate ways on the mainland," their nephew Ben Odom says. "But by the time they got out of debt they were sort of captured by the island."

They had also started building what Harry calls "the finest island store on the coast of Maine," a year-round operation with a walk-in cooler, butcher counter, and a remarkable array of luxury items. Monhegan hadn't had a store like this before, and hasn't since it closed more than a decade ago.

But Doug and Harry had another pursuit in winter: lobstering. They'd started fishing in dories, hauling lobster traps or tub trawling for halibut. In 1946 they bought their first lobsterboat, the JULIET H, and began fishing together. They would continue to do so for another half century. Their third boat, CHRIS, a wooden 30-footer, is still fishing out of Monhegan's exposed harbor.

But disaster struck in November 1963. One of the store's generators backfired, setting a grass fire that spread into the oil and bottled gas storage area. Exploding gas and 45-knot winds spread burning oil onto surrounding buildings. Doug and Harry had been building traps in the store when the fire started, and had nearly put out the fire when the well they were using ran dry. Town water had been shut off for the winter and hoses thrown in the sea became clogged with seaweed thrown up by the passing storm. "For a while it was feared the island would have to be evacuated," island resident Alta Ashley wrote at the time. Eight buildings were lost in the fire, including the store and Doug and Harry's home. "This island has been as if without a heart since the fire," Ashley wrote. [With islanders' help, the store was rebuilt.]

Today most islanders agree that the main threat to Monhegan's year-round community is the explosive growth in island property values. With modest seasonal cottages selling for half a million dollars, young fishermen find it difficult to purchase — or even rent — year-round accommodations. But over the past two decades, Doug and Harry have sold property to young fishing families for a fraction of what the market could have borne.

"I think Doug and Harry feel strongly that the island would be a different place without fishing families," says Tralice Bracy, whose husband, Robert, lobsters from Monhegan. The Bracys' new home stands on a lot sold to them by the Odoms at great discount. "If you're starting out as a young fisherman, you already have a huge investment in your boat and traps; and if you're investing in property as well, it's definitely a hard go. This was a huge gift for us from Doug and Harry and their family as well."

Harry Odom says Monhegan's unique community will persevere. With two-thirds of the island in the hands of the Monhegan Associates land trust and protected in its natural state, and a community of committed winter fishermen, Harry expects Monhegan's unique community will continue to persevere. "Fifty years from now," he says, "I don't think it will have changed too much. It's nice to have it that way."

2002

The Odom brothers, sketched by Jamie Wyeth in 2001.

CHAPTER THREE

Communities

"COMMUNITY" HAS BECOME SOMETHING OF A BUZZWORD for everything the twenty-first century is not: intimacy, familiarity, a sense of reciprocal responsibility, shared history and place. "You recognize your holding ground," writes Maine poet Philip Booth, "and know what deviation / swung your compass back." Community is thus about return — a going-back to a world that is more familiar and more appropriately scaled for human beings.

Island communities may sometimes be isolated, provincial, inbred, even bigoted against outsiders — but they also have inherent strengths that few other places in the contemporary world possess. "In the winter, I started an egg delivery route, dropping by my customers' houses once a week," recalls Chellie Pingree as she recounts her slow entrance into North Haven's tight-knit year-round community. "I came in through their kitchen doors and was occasionally asked to sit at their well-scrubbed or cluttered tables and drink coffee. Talking about the small things that fill our lives, I gradually began to know my neighbors and find out who was in this community…."

Communities don't have to be isolated rural places. Long Island in Casco Bay was a neighborhood of Portland, Maine's largest city, when its residents decided they had more in common with each other than the folks at City Hall — and declared their independence. Quite literally, they took control of their own future. As one resident put it, "We're finally taking ownership of our problems."

Small communities are always subject to large forces beyond their control. The year-round islands off the Maine coast face constant battles with the Maine State Ferry Service that is their lifeline to the mainland. Real estate prices and the rising taxes they bring with them are pushing many island communities in the direction of extinction.

David D. Platt

The Morris House (detail), N. C. Wyeth, 1935
Above: Peter Ralston

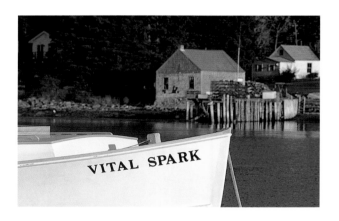

Community Design as if People Mattered

George Putz

I was a westerner born to the rationale of democratic plots of land, all in squared county sections and municipal blocks. The haphazard metes, bounds, and lays of properties in northern New England seemed like magic to me. It was a world created under Christmas trees for medieval imaginations, not reasonable land tenure. And so, on a smoky southwest day during the summer of 1963, when I first came to a Maine island and walked up a dooryard path to pick up a key, all my attitudes and perceptions about people in communities began to change forever. It would still be a few years before I thought of Maine island community layouts as much more than a wonderfully inspired accident, a kind of caprice allowed by rural isolation and cultural deprivation. Retrospective shudderings at that naive impression are only slightly relieved by the constant observation that all visitors to an island indulge the same romantic violence from the moment they step off whatever conveyance discharges them.

Domestic Maine islands grab our hearts for the sufficient reason that they are beautiful. Setting and proportion, architectural styles, and period design figure in our ordinary reasons, but these are easily celebrated, without intellectualization, by plain, unencumbered admiration. Much of it is period architecture

Frenchboro

PETER RALSTON(2)

Stonington

JEFF DWORSKY

without architects, design and siting from times without designers and planning boards, created by people who met their own requirements with heart, saw, and hammer.

Traditional Maine island neighborhoods are vibrant living places because they are wonderful spaces in which to play and work — indeed, to not make that distinction. Even when emulating middle-class 19th-century Victorian renderings (deluded) of 18th-century Aristocratic design concepts (depraved), island community builders had the individuality and good sense to regard each and every building plot without regard for rational planning schemes. Rather, structures were built to the characteristics of the site and to their own builders' eccentric preference. The result is an eclectic layout of structures and humane networks of inter-dwelling and outbuilding connections that ignore the block grids and sidewalks that organize and direct the ebb and flow of community intercourse elsewhere. This semi-secret sharing of paths through, around, and over the property of others is a continuing rite of intensification. Kids and adults

alike celebrate their citizenship by knowing all the ins and outs, the appropriate routes from here to there, and by using them. When these paths are abandoned and later forgotten, it is diagnostic of community exasperation. Island community death begins when too few remember where the paths went, or why.

Automobiles and a strictly cash economy have had much to do with the demise of the traditional, natural community traffic, but still there are contemporary children, and when they, too, fail to observe the ancient routes and byways, symptom becomes disease. In any case, it is the ungridded island community framework from which our notions here derive — matters of the sun, marine winds, harbors, horticulture, and husbandry, 19th-century maritime work, and the cantankerous wants of free people who had vital cultural memories of feudalism.

Island community design reflects the affective needs of people, not the imposed rational planning of experts or public administrators, whose milieu and work are essentially alien to community life.

(The irony here is that rational community planning was a child of the French Revolution, which took the old Roman and late Renaissance infatuation with geometric planning and declared it a means to perfect, non-elitist democratic community design; in the process creating inappropriate community layouts everywhere the system was applied, as well as the worst kind of elitism hidden under the guise of what's good for everybody.) In the corners of New England society, however, people individually negotiated their lives and living spaces and dealt with greed and stupidity in their own way, often in accommodating ways. On the Maine islands, these people created a unique tradition of American community design based on maritime requirements, an abundance of good construction timber, and, in retrospect, a competent Yankee contempt for pseudoscientific mandates and proprietary experts.

First is the sun: Whoever thinks that the sun is a rational and straightforward consideration has never looked at how sunlight actually falls on real estate. Astrophysics, geometry, the songs and dances of land planners and architects notwithstanding, the nuances and effects of the sun differ from place to place within very short distances, everywhere. Regardless of the architectural style –

Federalist, saltbox, cape, mansard, Victorian, etc. — the particular way that sunlight falls on and into a domicile or out-structure must reflect the special aspects of the sunlight on the building site. It was not just a matter of energy use. Even though our forbears always had heat energy in mind, many other considerations were more important – morning light in the kitchen, dryness in the anterooms, good midday reading and sewing light in the family and birthing rooms, western light in the parlor, decent winter light for the milch cow and buggy horse, and so on....

Then comes the wind: It does not always blow on the coast of Maine, but it does blow often. Winds have been an energy resource always, usually a nautical blessing, sometimes a curse or even a killer. In siting and orientating a new home or attendant structures, few men and no women ignored how the prevailing winds would serve or impinge on their ordinary lives. While acquiescing to the silvic wisdom that nearby trees and forest stands were not, could not, be permanent companions, our island forbears often planned on the care and replacement of tree growth in such a way that nearby land contours would influence the wind....

And always, the waterfront: The placement and orientation of homes and structures to the water

PETER RALSTON

Head Harbor, Isle au Haut

cannot be understood without a forthright vision of Maine coastal society — and in this respect mainland and island communities differ considerably....

The islands were the domain of private maritime entrepreneurship — owners and shareholding crewmen on relatively small vessels, whether in fishing or trade. Indigence was not a permanent social status. A few seasons of work would get anyone a few hundred dollars saved by, and that was enough to establish a home and household. Further-more, a view of the sea was a proud affirmation, an expressed relationship to the sources of one's life and livelihood. Island wives did not spend their best years beholden to absentee values and did not fear or loathe the sea on which their men worked, as did most of their mainland counterparts. They preferred practical, used and valued windows on the sea; not useless, hypocritical widow's walks and glazed cupolas.

JUDY GLICKMAN

These earnestly cherished views to the water did much to create aspects of island community design. Any island planning board that disallows this ancient consideration does itself badly. Maritime island life is sustained by small boats, nautical craft that do not require complex institutional paper or highly capitalized infrastructures (at least not until recently), and direct views of the

harbor go well beyond the aesthetic. Who's in? Who's out? Why all? Why not!? An island harbor was and remains not only its community's means to existence, but the symbol of each family's absolute citizenship. Harbors and approaches wanted watching, and watching windows were placed where they must, oftentimes orienting buildings out of linear or rational street alignment, even back from the water as much as several "blocks" or their equivalent for a larger view of the waterfront....

Almost everywhere on islands, just below the soil, you can find marbles – emmies left behind by eight-year-old children decades ago when the landscape was clear and in agriculture or grazing. Bless the eight-year-olds! May they return to create and remember pathways and byways that fix and define enfranchised and happy lives in Maine island communities. If their houses are not the old ones, may they be fine ones anyway, and create rooftops that fit no mainland pattern and express the human spirit, not imposed philosophies. Balance between loss and hope is impossible. Only love and spirit can steal loss's portended winnings. Resident islanders must recollect their spirit and cut through this madness between past and future.

1991

PETER RALSTON

PENOBSCOT

Children of the early
Countryside

Talk on the back stoops
Of that locked room
Of their birth

Which they cannot remember

In these small stony worlds
In the ocean

Like a core
of an antiquity

Non-classic, anti-classic, not the ocean
But the flat
Water on the harbor
Touching the stone

They stood on—

I think we will not breach the world
These small worlds least
Of all with secret names

Or unexpected phrases—

Penobscot

Half deserted, has an air
Of northern age, the rocks and pines

And the inlets of the sea
Shining among the islands

And these innocent
People
In their carpentered

Homes nailed
Against the weather—It is more primitive

Than I know
To live like this, to tinker
And to sleep

Near the birches
That shine in the moonlight

Distant
From the classic world—the north

Looks out from its rock
Bulging into the fields, wet flowers
Growing at its edges! it is a place its women

Love, which is the country's
Distinction—

The canoes in the forest
And the small prows of the fish boats
Off the coast in the dead of winter

That burns like a Tyger
In the night sky. One sees their homes and lawns
The pale wood houses

And the pale green
Terraced lawns.
'It brightens up into the branches

And against the buildings'
Early. That was earlier.

George Oppen, *1996*

A Place Slowly Earned

Chellie Pingree

In 1971 I stepped off a ferry and was on an island. A Minnesota girl, all of 17 years old, I had a long blond braid, wire-rimmmed glasses, and a metal-framed backpack containing all my earthly possessions. I had come to visit Charlie, a friend I'd met through a school program, who had taken up residence in an abandoned family cabin at the end of a mile-long dirt road. There was no running water, no electricity, and no TV. Armed with *Living the Good Life* by Helen and Scott Nearing, we wanted to go back to the land. During the summer we were surrounded by our friends from away who drifted through, shared our ritual diet of soybeans and brown rice, and talked about subsistence farming and forming an "alternative community." By fall they were gone, but I never left.

Charlie was from a summer family with a 100-year history, and when he didn't go home at the end

JEFF DWORSKY

of the season, people were confused. Although there were two or three families we visited sometimes, people rarely spoke to us in the store or asked us questions in the post office. We naively thought we were unnoticed, and it surprised us when we heard the rumor that we were spies for the summer people.

Charlie was able to find a job with the road contractor, but there were no jobs for me, so I stayed home making candles for a mainland craft store. During my second winter, a creeping need to be involved drove me to ask if I could volunteer at

the school. The kindergarten teacher "from away" was pleased to have my offer of help, and the principal had only to propose it to the school board for approval.

I will never forget the principal's expression when he came in and sat next to the woodstove in our two-room cabin. Reluctantly he told me there was a problem: the board had unanimously voted not to allow me in the school. He quoted one of the board members: "The girl who drives that red pickup truck is not coming in our school."

At the time I felt only bewilderment at these

feelings of animosity from people I didn't even know. What I didn't understand was that upon stepping onto the island I had entered a community. There were standards and requirements, unconsciously crafted by a community to ensure its longevity and stability. I hadn't even known what a community was, but I spent the next 20 years learning — and probably will never stop.

We moved away for three years, and in the separation people seemed to warm to us a little. When we returned for summer visits, people would ask us when we were moving back. Although we fully intended to start a new life on the mainland — we even built ourselves a house there — the pull of the island was strong. We wanted to come back. We did — married, with our first child, armed with a college education and vocational training — determined to fit in.

We chose work that suited island life: I ran a farm (cows, chickens, sheep, and vegetables), and Charlie built boats. In the winter, I started an egg delivery route, dropping by my customers' houses once a week. I came in through their kitchen doors and was occasionally asked to sit at their well-scrubbed or cluttered tables and drink coffee. Talking about the small things that fill our lives, I gradually began to know my neighbors and find out who was in this community. I heard about their mother's health, how to cook salt fish, who was related to whom. Then I went home and scrubbed my table, did my chores, and retold the stories to Charlie.

Every day people dropped by to pick up their glass jugs full of rich yellow milk layered into skim and cream. I invited them into my kitchen and was honored if they accepted. I poured them coffee, proud of the clean glass milk jugs draining in the sink, and I'd ask them questions until they were tired. Gardening was a favorite topic of conversation: I learned to start my tomatoes on town meeting day, not to touch my beans after a rain, and how to fool the slugs. People loaned me their butter molds and told me how their mother got the last drop of buttermilk out of the rich yellow grease. I learned to quilt, I learned to knit. Before long I was the mother of two babies, then three. Sometimes a few other young mothers would come by, and we'd watch each other's children and exchange stories, wondering if we were good mothers. Older women would drop by too; they told me about rubbing vodka into a child's emerging tooth, staying away from Vaseline for diaper rash, and rubbing Avon crèmes into my own face to postpone the wrinkles that were sure to come.

In July and August I sold vegetables to the summer people on Mondays, Wednesdays, and Fridays from the shed attached to my house. Sometimes a customer showed up on a Saturday,

pleading for tomatoes for a critical dinner party, or called long after I'd gone to sleep, asking me to save the highly desired but rationed sweet corn for her the next morning. I began to understand the price my island neighbors paid to earn their living. Summer exhausted everyone; winter was when the community reconnected.

We raised pigs and they escaped from every pen and we chased them through our muddy pond and neighbors' fields. We bred a sow once, who rested on the backseat of my VW van as a friend and I took her on the ferry to meet her mainland mate. The sow returned a month later and got larger by the day. Shortly before the piglets were due, she became hopelessly lame. When no drugs or treatments would bring her back to her feet, we consulted the man in town who knew the most about animals — the chairman of the school board. He came over and shot her. Then he showed us how to rub her with tar and hoist her with the tractor bucket to dip her body in boiling water, so that we could scrape her hair off the hot skin. Finally we cut a slit down her middle and stood back while her hot insides and never-to-breathe piglets tumbled out.

We turned to him for advice on other occasions after that — how to hobble a cow who wouldn't keep her foot out of the milk bucket, how to treat a calf with diarrhea. I hadn't become an islander, but I was learning to ask questions.

People helped us with haying and we ate a big meal together when it was done. Friends came over when we plucked chickens or froze sweet corn and we shared the results. The more people learned about us, the more often they would speak to us on the long ferry rides, ask us along on Sunday picnics, invite us over for dinner. I learned a paradox of island life: to be accepted, or even given your privacy, you had to let people know a little about who you were. They had to know who they were allowing to join the community, whether you wanted to participate or not.

We began to go to meetings where the community plotted its course and made everyday decisions — town meetings where salaries were set for road crew workers and fire trucks were talked about and the recreation council defended the money for an outdoor basketball court; school board meetings where parents disagreed with teachers or principals and told them what they thought. At first I was silent, rehearsing the words in my head before I spoke, even though they were mild. Eventually I even learned to say things that others would disagree with, about taxes or schools or the ferry schedule. I learned that people could get angry at me in a meeting and the next day in the store would ask how my wood was holding up this winter or whether I would be plowing gardens

Fox Islands Thorofare

PETER RALSTON

in the spring. I learned that on an island people depend on each other too much to hold grudges over any but the most important disagreements.

I rode the ferry, never knowing who I was going to sit next to. Often I found myself in conversations with people who had disagreed with an article I'd written for the local newspaper or who didn't like me because I was from away. We'd end up talking about which potholes in the road were hardest to avoid or how many trees were lost in the last storm or maybe how sad it was to have a relative die. As the boat docked, we'd both be thinking about our commonalities.

Sometimes the ride was rough. If there were children along, mothers would shuffle through their canvas bags looking for something for their children to get sick in. If the mother was sick, someone else would comfort the child while she went out on deck for fresh air. On one gray day, a northeast wind was building at boat time. The captain hesitated before starting the engines for home, but once the ferry was loaded, he set her out to test the waters and then he just kept going. The boat pitched and trembled in spite of the captain's best efforts to tackle each wave gently. An older woman sitting on a hard plastic seat was nervous, and everyone's attention gradually focused on her. The cabin, with seats for 30, became cozy and enclosed us all tightly as we huddled in the back. People told stories about other storm crossings and joked with her. A young man whom I had watched grow up leaped up to share her seat, teasing her not to tell his wife. She laughed and blushed and for a moment didn't think about the next shudder and crash that was soon to come. It was almost an embrace — a small, random group of community members helping her through the ride and giving us all comfort.

I was two seats ahead of the old woman. In the seat between us was her daughter, a middle-aged woman, once a member of the school board that kept me out. She was not my valued friend. I had shared many ferry conversations with her and listened to her wisdom about children, the community, and the school board, of which I was not the chairperson. It had taken 20 years, but in spite of how little I knew when I made that first step, I was now part of this random group. I had never created the alternative community of chosen peers that my friends and I had once discussed. But here on and island, I had come to belong.

1992

The Little Island
That Got Away

Al Diamon

Almost unnoticed, the Long Island Research Committee was formed in July 1991 to investigate whether secession from Portland made sense. The chairman of the committee was Mark Greene, a part-time island resident who taught school in Massachusetts. At the same time, the Long Island Civic Association, headed by summer resident Christine McDuffie, a substitute teacher from Portland, began negotiating with the city for more services.

Portland was making a clumsy effort to counteract the rebellious mood on the islands. It released a study showing many island homes were selling for far more than their newly appraised values. The state tax assessor reviewed the revaluation and concluded it was fair. The assessor even emphasized that most island properties were still undervalued by 10 percent or more. Since most islanders didn't want to sell, these statistics provided cold comfort.

August 29, 1991, turned out to be a turning point for the secession movement. City officials took their annual tour of the islands that day, and they came equipped with some ideas to ease the rebellious mood. At each stop, councilors talked about setting up a special tax district on the islands where lower

JOHN PATRIQUIN

Other Casco Bay islands' moves towards independence were thwarted by Portland.

property tax rates would apply. They also discussed making the islands into village corporations within the city, which would allow them to keep more of the tax revenue they generated. Since Portland collected about twice as much from the islands as it spent on them, the idea had definite appeal.

The islanders on Peaks, the Diamonds, and Cliff were interested. Secession movements on the islands were going nowhere, and most of the leaders were willing to settle for tax relief instead. "Secession is a long-term, separate thing," said Peaks Island's Charlotte Scot. "The majority of islanders are worried about their tax bills now."

When the officials got to Long Island, McDuffie had a different message. She handed councilors a memo thanking them for their help in removing junked cars, improving recreational facilities, and maintaining staffing on the city's fire boat. But the memo also listed nearly two dozen unresolved problems, and warned that the revaluation had been "wrenching" and "too much for many in our community to absorb."

"We feel we have no choice but to begin to investigate our other options," she wrote. "The local control of wrestling with our own problems and setting our own priorities could perhaps make

us more resilient to the threats that face us as one of the few remaining year-round island communities."

McDuffie's quiet message didn't penetrate. Councilors sipped beers and played penny ante poker as they sailed back to the mainland. They were convinced the secession movement was dead and the tax revolt was waning. Even if McDuffie's little group wasn't satisfied, city officials figured Long Island was too small to go it alone.

Nobody on board seemed to realize they had just blown their last chance to derail the secession movement.

Long Island had been part of Portland since the city split from Falmouth in 1786. This island is 4.5 miles from the mainland. Its area is just 900 acres, its length only 2.6 miles, and its width less than a mile at the widest point. Its year-round population is estimated at 175, swelling to nearly a thousand in the summer. It has one store, one seasonal restaurant, no hotels or bed and breakfasts, and no businesses except fishing and lobstering. Its only tangible asset is property, which the revaluation showed had increased in value more than threefold, from $11,228,660 to $37,455,300.

In the minds of Portland city councilors, Long Island had none of the elements needed to operate independently. The idea of a special tax district for the islands gathered dust. It was June 1992, before the Portland City Council even got around to formally rejecting the idea. Opponents of the tax district argued it wouldn't be fair to mainland taxpayers. City councilor and Little Diamond Island resident Ted Rand warned, "If you turn this down, it's bye-bye."

But by then, the momentum toward secession was probably unstoppable. On October 20, 1991, a community meeting on Long Island voted 98–2 in favor of holding a referendum on forming a separate town. The next day State Representative Ann Rand of Portland introduced legislation to allow a secession vote in November 1992.

The islanders studied hard. They met with other island municipalities to learn how they handled the nuts and bolts of governing. They talked with the Cumberland County Sheriff's Department about the costs of police protection. They drafted a budget that called for an annual tax rate of $15 per thousand, about a third less than Portland's rate. "Paying smaller taxes is the prime impetus," said McDuffie.

That "prime impetus" was severely threatened in January 1992, when Portland announced that if Long Island left the city, islanders would have to pay their share of long-term debt accumulated while they were part of the city. The secessionists' draft budget estimated that cost at $130,000. City Manager Bob Ganley's estimate was $750,000. The city also wanted Long Island to bear the cost of closing its dump, and warned it would not give up any assets, other than property, if the island became a separate town. If Portland had its way, most of the tax savings would be eaten up for years to come.

Supporters of a separate town were quick to react to the threat. If there were no savings, the central point of their campaign was gone. They needed a new theme, and by the time the Legislature's State and Local Government Committee held a public hearing on the secession bill in February 1992, they had one. "This is not about a tax revolt," Greene told the committee, "this is about preserving an endangered way of life."

The pro-secession group also cast the battle in David-versus-Goliath terms. Long Island and Portland were "a total mismatch," according to islander Robert Jordan. "We have no political impact on the city; they don't even know we exist."

Portland's counterattack was unfocused and ineffective. City Councilor Clenott asked, "Why are we being pushed into making a decision in a matter of months on a relationship that has endured for 200 years?" Councilor Tom Allen argued that Long Island, with fewer than 200 residents, was too small to be a separate town. That got laughs from legislators who represented towns half that size.

Most Maine legislators look on Portland as an alien place, full of people from away and flatlander values. Consequently, rural representatives relish any opportunity to stick it to the state's largest city. They weren't going to require much persuading that letting Long Island secede was the political equivalent of freeing a virgin from Sodom and Gomorrah. On March 4, 1992, the committee voted 13–0 in favor of the bill. During House debate a week later, Portland State Representative Fred Richardson warned, "This is the beginning of a tragedy in the form of the emasculation of the City of Portland." The House approved the operation 128–7. The Senate gave its unanimous approval without debate or a recorded vote....

Technical problems held up final passage of the secession referendum bill, but on March 23 the measure flew through both chambers with only three dissenting votes. The governor signed it, setting the vote for November 3....

On November 3, 1992, Long Islanders went to the polls in record numbers to support secession in a 129–44 landslide.

The winners didn't spend much time celebrating. They had a lot of work to do to prepare for independence on July 1, 1993. And they had no illusions about how difficult it would be. "We're looking at three to five years of hell," islander Cynthia Steeves told the Associated Press. "We're willing to stay the course."

The first step was to elect some leaders. At a December 12 town meeting, secession leader Greene was chosen as moderator, but most of the other elected positions went to full-time residents with deep roots. Volunteers fixed up the town hall, helped draft a budget and sold T-shirts, which constituted the government-in-waiting's only form of income until tax revenue started flowing.

An official town meeting in May approved a budget of $683,000. Not a single attempt to cut an item won. Several efforts to increase spending succeeded, adding about $10,000 to the total. So much for the tax revolt. More than 100 people took part in the meeting, and the most contentious issue was deciding how many streetlights the town could afford to operate. It was settled by compromise. So much for feuds.

Not everything went smoothly. Islanders were cited for illegally pouring a concrete slab for the new town's transfer station without proper permits. Portland Public Works employees stripped a truck of required safety equipment before turning it over to Long Islanders. When the truck was sent to Portland on a barge to pick up supplies, city officials were waiting to ticket it. Portland and Long

PETR RALSTON

Casco Bay

Island workers got into a shouting match on the public pier over which municipality owned a lawn-mower. A summer resident was elected to the board of appeals, an apparent violation of Long Island's newly adopted municipal ordinances.

On July 1, 1993, the chalkboard outside the island's only restaurant, The Spar, listed only one daily special, "The Secession Sandwich." It read: "Let us alone." But there were few other signs of bitterness as assembled dignitaries officially incor-porated the Town of Long Island. The crowd even managed to give Portland Mayor Anne Pringle, a secession opponent, a warm greeting....

Down at the municipal pier, the cars are regis-tered and the public drinkers are few. There's also a new rope handrail to help ferry passengers climb the icy, wind-buffeted hill. It was installed by volunteers using donated materials.

"It's a practical solution that just appeared," said Chris McDuffie. "We're finally taking ownership of our problems."

1994

Living There

Cynthia Bourgeault

I t is simple and it is precious. Some are born to it. For others, it is an identity consciously chosen and, in some cases, fought hard for and won — the right to call oneself an "islander." Strictly speaking, the term "intentional community" applies to a place like a commune, a monastery, or an enclosed culture such as that of the Amish, where people willfully come together with the hope of proclaiming and living out some set of values that are crucially important to them. While it may at first seem far-fetched to apply this category to island communities, on closer inspection the criteria hold up surprisingly well — and in the process offer a telling commentary on where we members of the larger, end-of-century American society are in our effort to proclaim, choose, and live out the things that are truly precious to us.

From the first time I set foot on Swan's Island in 1972, I could feel it, almost welling up through my shoes: the call to a life that seemed wildly more authentic, more true, more compelling, than anything I could live back home in Philadelphia. I came full-time in 1979, and for something more than a dozen years Swan's Island was home. I lived, sailed my boat, built a home, raised a family, scrounged for a living — house painting, mowing lawns, teaching on the mainland, whatever it took. Every moment was precious. Still is.

For the past three years, I have lived in Colorado, next door to a more classically "intentional com-

PETER RALSTON

Crew Neck, Andrew Wyeth, 1992

munity" — Saint Benedict's Monastery, a small, Trappist contemplative community making its living through ranching and meditation retreats....

An intentional community in the classic sense has finite boundaries. Here at the monastery they call it "the enclosure." You know when you're in it and when you're out of it. It is the geographical space in which the rules and values of the community hold sway.

A lot of people point out, quite rightly in some ways, that island communities are really not that different from small, isolated mainland communities: still populated by traditional New England Yankees with similar values of frugality, reticence and self-reliance. One of my mainland friends insists that Vinalhaven is simply "a little Belfast miles out to sea."

True, but there is that one all-important distinction: an island is surrounded by water. It has finite boundaries.... It is this element of finiteness that puts the distinctive intentional spin on the otherwise generic similarity to stock New England community life....

In all my years on Swan's, I never locked my home, and the keys stayed in the car. Our one local robbery was an island classic. Once when a neighbor, a fine-art collector, went away for a while,

thieves came from off-island, broke into his house, and looted his collection. The sight of an unfamiliar van headed down his remote road attracted a neighbor en route, who went over to check out his house for signs of mischief while the van waited in line for the 1:30 boat. When the van rolled off the ferry in Bass Harbor, the state police were waiting to meet it. In a world where crime and violence seem to thrive on anonymity, that rootless world where nobody knows one's neighbors or cares, an island seems to be, and still remarkably is, a place where the high end of finiteness is a safety too rare in the world at large.

Deeper than fear, however, the driving motivation for many who come to islands, I believe, is the quest to find one's identity, a place where one belongs, with skills one can contribute. This sense of identity has both a personal dimension and a community one.

On the personal level, a good deal of the attraction of island life lies in discovering and developing skills in oneself: skills of self-reliance, ingenuity, that sense of being actually and personally in control of one's life....

I arrived on Swan's Island with a Ph.D. and ten years of college teaching under my belt. But medievalists were not highly in demand on Swan's,

Sons of fishermen, Metinic Island

so I started my working career bagging scallops at the co-op, then went on to mowing lawns and painting houses. Along the way, I learned precious skills: I could saw, hammer, replace soffits; could fix my car when it wouldn't start, sharpen a chain saw, re-putty a window, frame in a wall, lay a foundation — practical skills that made me feel connected and useful in a real universe. The pride I felt in my own self-sufficiency is more than just a vestige of my old tomboy spirit; it speaks, I believe, to a basic yearning in all of us to see the relationship between means and ends, to draw satisfaction from the work of our own hands and from living up to our uniquely human responsibility truly to tend a small piece of the earth that has been entrusted to us.

Perhaps the most powerful magnet drawing people to island life is the tangible sense of belonging to a community, a finite universe where folks watch out and care for one another....

And reciprocally, this finite universe is a place to share one's gifts. Many who have come to island communities have made generous donations of their time and resources: founding libraries and historical societies, serving on boards, helping out at the school, sometimes taking key roles in island conservation or economic development initiatives....

For me, there was always a more philosophical dimension to this as well. Knowing your place in community also meant watching the progression through time. The kid whose mother I tutored when she was pregnant is now old enough to buy my house; several others whom I watched enter kindergarten are now homeowners and civic leaders.

The native-born islander comes with a birthright as special as any ever to be conveyed, something tantamount to an inviolable place, unconditional acceptance. Island tots are community property, really; everyone knows their names, says, "Hello, deah," and fusses over them at the store. Only by really seriously "being a jerk" is that birthright ever jeopardized; and even then, one remains a native son or daughter.

For those coming "from away," acceptance must be earned; a place in the community must be won. Some newcomers, I notice, wear this very lightly, while for others it becomes almost a grail quest to have the mythical "they" confer this sacrament of belonging to "us."

Whatever form this takes — that initial breathless excitement or idealism that casts one up on the beaches of an island — the acceptance gets under way in earnest with the first making it through the winter. Those first couple of years will test the depth, realism and resiliency of the original commitment. As the days grow short, the weather raw, the winter endless, as tempers fray and distractions dwindle, the would-be islander will come face-to-face with the reality of island life and with his or her own internal resources.

A would-be islander, like a novice monk, may place himself under the care of a wise elder, and, if he is alert and adaptable, will catch that he is being molded to a basic pattern that allows for survival in close community. He will learn to talk very little, to listen first and hard, not to ventilate his opinions, to be alert to gesture and innuendo, and to submit his own self-interest to the greater needs of the community. These are not only virtues in their own rights, but survival skills for both the individual and the community.

It is a mixed bag emotionally. Anything worthwhile and difficult always is. But as I listen to the stories, the words, the hopes, of those who have stayed and those who have left, I feel the saga of something very moving being attempted — something very humanly significant. From these many folks who have voted to cast their allegiance with a small scrap of earth in the sea and the folks already floating on it, there are lessons to be learned of import to our wider human culture.

The most powerful lesson is the relationship between scale and human value. In contrast to the predominant culture, which tells us that the meaning of life is to get, to satisfy, to consume, to possess, island community says that real meaning is to find a place, to put down roots, to participate, to serve. If the culture tells us our goal is to be fully an individual, to grow to the max, to "do our own thing," island culture suggests that the only sustaining identity is ultimately relational — in a community, in time — and that growth is not by maximizing, but by pruning, like the strawberry vine, so that real blossoms and fruits emerge. And in contrast to the sense of self formed in a society that is fast-paced, crowded and sprawling, island identity is a fierce, tiny particularity, grounded in a finiteness, surrounded by solitude and face-to-face with the rugged force and beauty of nature, that reminds us of our true place in the scale of things.

For whatever shortcomings, living on Swan's Island was the most profound and formative experience of my life, and wherever I may be, I will never leave it behind.

1997

PETER RALSTON(2)

Fifield's general store, Vinalhaven

An "Unpretentious Exercise"

David D. Platt

A summer weekend in the mid-1970s, Charles McLane placed telephone calls in search of three elderly former islanders, in hopes of arranging interviews with each. Their recollections, he knew, would add to his understanding of Maine island communities, whose history he planned to research and publish when he retired from the political science faculty at Dartmouth College.

The phone calls brought disappointment. All three of the hoped-for informants had died within the past month; their knowledge of and insights into a particular way of life were lost. Losses can become lessons, however, and this one would have a positive effect on McLane's island-history research project. "At a certain point," he wrote in the preface to his first volume of *Islands of the Mid-Maine Coast*, published in 1982, "the idea of piecing together the history of these islands acquired an urgency" — because the living record, at least, was slipping away with the lives of the islands' older residents; because McLane was himself near retirement age and facing a project he knew would take him years to complete. "The study," he had realized by 1982, "would not wait until the illusory 'retirement.' It had to be done now."

What McLane and his wife, Carol, called their "unpretentious exercise" for retirement quickly evolved into something much more ambitious: a definitive history of three centuries of island habitation and use, covering hundreds of islands that lie along the Maine coast, all the way from the Kennebec River to Machias Bay.

The project took 20 years to complete. Undeterred by the deaths of those three former islanders they had hoped to interview in the 1970s, the McLanes went on to speak with scores of other island elders, incorporating their memories into the histories of the islands where they or their families had lived. More significant in terms of historical technique, however, was the McLanes' recognition that much valuable material lay forgotten in the records of towns and counties along the Maine coast — registries of deeds, court and tax records, town meeting reports — and in the statistics gathered each decade since 1790 for the U.S. Census.

History, like much of science, is largely an exercise in focus and perspective. How one looks at the available facts — how one assembles and analyzes them — is fully as important as the facts themselves. What Charles and Carol Evarts McLane learned through interviews, or ferreted out of family records and registries of deeds, wasn't particularly startling, revealing or surprising; what was new was the way the McLanes used the available facts to tell a story that had not been told before....

After 20 years of research, Charles McLane can generalize, just a little. "Islanders," he says, "have an innate capacity to submerge their differences in order to get along." The success of an island community "depends on who you're dealing with — the chemistry of who is on an island at what age, vitality, health, makes a lot of difference. What makes for success or failure on an island would be a damned good study...."

1998

COURTESY OF CHARLES B. McLANE (2)

A Beneficent Magic

Philip Conkling

Where to start? At the end or the beginning of John Wulp's story? In a sense it does not much matter, because you could say that his beginnings and his endings are circular — arriving where they started for a stageman, theater director and artist, who washed ashore by accident on a Maine island more than a decade ago. Once there, he discovered the island's community and became the greatest teacher of drama to school children that the state, perhaps, has ever seen. At the North Haven school, John Wulp has achieved unimaginable successes with a drama program in a small and insular place, with its gifts for mimicry. He created not just a powerful new vision within his adopted island community, but re-created himself, one senses, in the process....

Wulp: I always ask myself, "Why did I come here?" I've never understood it. We were on this boat trying to find Great Spruce Head Island, but landed in Pulpit Harbor. As soon as I landed there I really had the oddest sensation. I thought, "Oh my God, I'm going to live here," which is really very odd, to have that strong a feeling the minute you step into a place. So that fall I came back to find a house on North Haven. I had a very clear image of the house I wanted, but I found it instead across the Thorofare on Vinalhaven.

...I came here to Maine out of a sort of desperation, really. I owned this house, but I was broke. I had no money and I had no very clear idea how I was going to support myself, but I had nowhere else to go.

The setting: North Haven Schoolhouse near the center of the island. It is a redwood-sided building housing grades K–12, where some 60 island children scattered across 13 grades are clustered in multi-age groupings, once common in one-room schools. As in most island communities, the school is the central institution of community life, where all its best hopes and worst fears come careening into view. Barney Hallowell, the principal, originally from a summer family, has worked in the North Haven schools for over 20 years. The origin of the island's drama program, he explains, resulted from observing many island kids' well-honed talent for mimicking their elders, often in wickedly funny ways. Hallowell recognized that his school was too small to excel in sports, but for a school to become great, it has to be good at something, anything; he

BRIDGET BESAW GORMAN

sensed that a drama program might be the ticket.

Wulp: When Barney Hallowell asked me if I would teach, in a way I grabbed at it, because I thought it was a way of improving my lot economically. I was working all the time. I worked at the lobster plant packing lobsters and I worked as a cook downtown when Phil Crossman had the Crow's Nest, because there was nowhere else to go....

The setting: North Haven Community School at a school board meeting shortly after town meeting election, in which the balance of power on the school board tilted to a majority opposed to the direction of North Haven's school program. In

addition to the basic curriculum, this program provided for off-island school trips and an arts and enrichment program.

Wulp: Then we did *A Midsummer Night's Dream*: just getting it on was a big success. What happened? Well, it took about an hour and 45 minutes, whereas usually it takes about two and a half to three hours to put on. It moved with a speed and a clarity that was unusual. Again, it was a matter of a clear articulation of the text. Just say it, let the play be heard. And it was a great success....

Now, nearly every kid in the school, except those whose parents won't allow them to, has participated. I would say 90 percent of the school has been in plays of one sort or another. And one year, we won the regional one-act play competition — and another, the state one-act competition. We've done very, very difficult material: Shakespeare, Anouilh, Thornton Wilder and Oscar Wilde. Most of the plays depend on language and the kids have learned to do this in a forthright sort of way. Certainly a study of the sort of plays we do would be demanded in any school with high academic standards.

I would like the kids to regard the theater as something holy and sacred, and that when they perform it's like this interchange between them and the audience that is, in strange ways, divine. I mean what is theater? We come together and for a moment we are joined as a community ... watching this thing in front of us that somehow defines our lives. Theater on the highest level is some sort of mystical experience. It's one of the few times when we become omniscient, like gods. You know, we watch this experience and we can understand it. We can never do that really in actual life....

The setting: North Haven Community Building, a short walk up the hill from the ferry landing. It's early April, but winter's chill clings to the water like a dark cloak. North Haven has struggled through a devastating community schism for

two long years, pitting neighbor against neighbor and family against family. Recent elections, for the school board, and resignations, among the selectmen who opposed the direction of island education that the drama program came to represent, have changed the balance of power on the island to the status quo ante. But everyone knows nothing has really changed the deep dynamics of the underlying conflict, and maybe nothing ever will.

Into this deeply disturbing situation comes the school production of *The Wind in the Willows*, a whimsical musical with a cast and production crew of over 50 kids drawn not just from North Haven, but also from Vinalhaven and Green's Island, to stage the show. A standing-room-only crowd of

over 150 people is crammed into the rows of chairs and bleachers. Down front is a delegation from the New York theater world, including the playwright, Arnold Weinstein, whose play, although written a decade ago, is being performed for the first time here tonight.

Wulp: *The Wind in the Willows* was quite a step for us because it was the first time we did an original play, that anyone entrusted us with an original play. It had never been performed before. We just did it and it worked. The kids were amazing. We used a whole new bunch of kids; we used mostly people from the lower grades. We still had to use some of the older kids like Asa and Chris Brown, but mostly they were kids just coming along.

When you see kids up there on the stage you cannot believe the kids are not getting a good education. But on the other hand, I have never been altogether able to do what I hoped — to give them enough confidence to feel that they could go anywhere and do anything they wanted to do. I think there is a fear, a certain naiveté, that is difficult to overcome. They are innocent of the world.

The cast of *The Wind in the Willows* is put together with a keen sense of the mythic identities that lie just beneath the surface of island kids, seemingly just waiting for a moment like this, for a director like this, to create the space and unlock their hearts. Ratty is as lyrical as the Mole is earnest. Badger is a high school youth who can seem to fill a door frame but who has not been on stage before Wulp came to North Haven. He is wonderfully solemn and ponderous in his portrayal of the ultimate enforcer among animals both good and bad. The slinking, black-costumed weasels and stoats are horrifically wonderful in their roles. The chorus of mice, with the heart-stopping solo performed by a flaxen-haired first grader, nearly brings the house to tears. But the show is stolen by the portrayal of the spectacular excesses of Toad, played by Jacob Greenlaw, a quiet eighth grader who works in the island grocery store. Jacob, like the play itself, is new to acting before a crowd of friends and island neighbors, new to costumery and lights, new to it all.

The production of *The Wind in the Willows* is an extraordinary display of the power of the arts to drive our worst demons back into the shadows, if only temporarily. Where the limitless energy of such a production comes from is unknowable, but the ability to find new meanings in such familiar territory is a kind of transformation. We search for this transformation in the sacred space of our lost childhood that we imagine, if only for a moment, we can recapture.

1999

BRIDGET BESAW GORMAN (2)

The Mr. Wulp Effect

Karen Roberts Jackson

Nearly two years ago my two teenage sons, and thus, our family, became involved with an enigmatic man by the name of John Wulp. He, at that time, had a long list of adjectives, and a few expletives, trailing along with his name. He is an eccentric, demanding, omnipotent, retired New York City director who stages jaw-dropping performances by kids on North Haven island. We heard rumors about him on both sides of the spectrum: he broke kids' spirits, he asked too much, he was impatient; mostly, he was intolerant of anything less than perfection. On the other side, lo and behold, he was extracting "perfection" out of junior high and high school students.

Our family also had a few oddities and adjectives of our own, such as the fact that we homeschooled our four kids on a small outer island — Green's Island, half a mile or so off Vinalhaven. I remember that we met Mr. Wulp (as all of the children call him) at a poetry reading at his home, Stone Farm, a sanctuary of art and literature. I also remember that in the kitchen was a table laden with the most mouth-watering, knee-buckling, ambrosia-of-the-gods desserts, created by this same quirky fellow. Sitting on his couch making small talk, I recall the woman next to me taking down a trophy from the mantelpiece and whispering to me, "Who the heck is Tony?" We were later to learn that, along with his "Tony," he had received an Obie and actually a long string of Broadway-related awards that impressed me, though I was fairly ignorant of their true significance at the time. Soon it came about that my two sons would work with him on his next production, An Evening of Shakespeare: Sonnets and Soliloquies....

I'm fairly certain that anyone who has had a relationship with Mr. Wulp would consider it unique. He extracts, demands, conjures a unique response, emotion, involvement with each person he meets. For better or for worse, he draws forth a full response from you, "casts" you, demands from you that which you might not have offered otherwise. I watched him draw forth a devotion and concern from my sons that I would not have thought them capable of in that phase of their youth. He could get them to cut, or not cut, their hair in the fashion *he* prescribed. He would get my

rebel sockless and shoeless son to wear shoes and *white* socks by calling him up at 6:30 a.m. to remind him. He could get the boys to take out their earrings and the girls to vamp about convincingly with feather boas, circa 1920. He could demand that they give up not only their Saturday night carousing, but their Sunday afternoon as well. In some mysterious way he brought a discipline to their lives that a mother would die for.

Mostly, he expected it of them, and they came through for him....

1999

BRIDGET BESAW GORMAN

PETER RALSTON

from The Country of the Pointed Firs

The dark spruce woods began to climb the top of the hill and cover the seaward slopes of the island. There was just room for the small farm and the forest; we looked down at the fish-house and its rough sheds, and the weirs stretching far out into the water. As we looked upward, the tops of the firs came sharp against the blue sky. There was a great stretch of rough pasture and here were all the thickly scattered gray rocks that kept their places, and the gray backs of many sheep that forever wandered and fed on the thin sweet pasturage that fringed the ledges and made soft hollows and strips of green turf like growing velvet. I could see the rich green bayberry bushes here and there, where the rocks made room. The air was very sweet; one could not help wishing to be a citizen of such a complete and tiny continent and home of fisher-folk.

Sarah Orne Jewett
Island Journal, 1985

PETER RALSTON

Island Fellow, Emily Graham, with student on North Haven

Island Fellows

An Island Institute program inaugurated in 2000 matches recent college graduates with pressing community priorities, placing them as "fellows" in residence at the invitation of island communities. In 2001, ten Island Fellows were living on islands from Casco Bay to the Mt. Desert area. They functioned as teachers, fisheries researchers, coaches, consultants and volunteers in a variety of community projects. Most important, they did their best to become full-fledged members of their adopted communities. All were encouraged to write about their experiences, and what follows is a sampling of thoughts from the 2000–2001 group.

THE MEANING OF THE DARKNESS ON
NORTH HAVEN
Emily Graham

Winter has come to North Haven. I like the phrase that winter has come; it implies that winter is something that exists — something that visits for a while. It's true, and it's our turn to have winter as a guest for a while. It seems to be settling in, too: we got about 4 inches of snow last night and this morning.

Taking the ferry back to North Haven this evening was beautiful. Coming in to the island the water was like a mirror, completely still, reflecting the boats and the lights of the little downtown area. Before coming into the ferry landing and town, though, there were very few lights. I knew that there were houses all along the shore there; I wondered why. When one of the students came by, I asked him. No lights because the houses all belong to summer people. The darkness was so striking to me — a real, physical manifestation of that sharp divide between those who are here year-round and those who just come for the summer. It all sort of came together for me — a first real glimpse of understanding North Haven and the seasons of people who live here....

TEACHING CONTEXT ON
VINALHAVEN
Mike Felton

One set of fraternal twins, one set of identical twin girls, amazing writers, gifted painters, an athlete who is already a junior legend in the high school and elementary school, a redhead whose comments drive Mom and teachers to fits of frustration and bouts of laughing, a boy who one week is present but really not there and the next is working diligently to complete his work, a girl who has a maturity and depth of thinking that might propel her to any of the nation's best universities or colleges. These are only a few of the personalities who combine to form the Vinalhaven 7th grade. It is our responsibility to form a class in which we work together so that we both might succeed; themselves as students and myself as a teacher.

History is more complete when we include more of the context, causes, and origin of events, movements and peoples. These events, movements and people are central parts of American history. To ignore them is to ignore the truth and

to be blind to the whole story. Serious students of history need to see both the darkness and the possible light emerging from that darkness. To not recognize the darkness prevents one from being able to appreciate where the light emerged from, or the context from which the light shines. Moreover, students need to focus on the tragedies of American history so that they might address possible continuations of those tragedies in their nation's society today....

AS THE WORLD TURNS ON
PEAKS ISLAND
James Essex

Here comes the day, another spin into the path of the Sun's rays. Today is a little different — ice coats the trees concentrating and refracting the light like a million different suns, a beautiful show. Ocean is calm. Atmosphere is calm. As suspected, the thermometer is broken, it shouldn't be 10 degrees F with the sun already turned on. I push on the door in a vain attempt to nudge the Earth's tilt back in a favorable orientation, and imagine stealing a few moments of summer from the people on the south half of the planet. A tough Maine raven, too proud to retreat south, breaks silence as it lifts from a

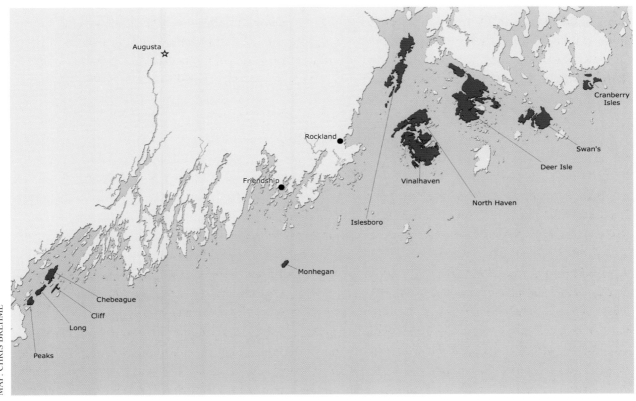

MAP: CHRIS BREHME

tree perch; the branch flexes and the ice cracks. The raven has cracked today's whip. Time to work — not outside today, the winter solstice — for no other reason than theoretically it should be the coldest of the year. This day, like the previous 99 I've spent on Peaks, will be devoted to developing a working knowledge and record of the spatial and temporal distribution of matter and energy on Peaks Island. The mix of technology used to catalog and define where things are and when things happen entertain me into the night.

There goes the day, just like it started: calm ocean, calm atmosphere, broken thermometer.

AN ENERGY CRISIS ON
ISLESBORO
Kathleen Reardon

Dec. 18:

Here I am writing by candlelight. It wasn't snow, but it was one hell of a storm. From the rain that started just as I came home Saturday night till mid-afternoon today (Monday) when the cold front finally cleared things out — the weather has been insane. It poured all Saturday night, raining in sheets that pounded against my windows. And with that rain came wind, lots of it. The weather reports said normal gusts of 35 to 40 with some extreme gusts of 50 to 60. All day Sunday the winds blew. Every once in a while the power would flicker, but it didn't go out. Mid-Sunday, I drove down to the Town Beach at the southern end of the island to see what the water was doing. Although nothing like it probably was off Vinalhaven or Isle au Haut (what I would have given to see that!), the waves seemed as if they were eating the shore. The power of the wind and waves was amazing.

Now it is Monday. My power didn't go out until early this morning. I woke up around 6 a.m. to a branch beating continually against one of the skylights, and to darkness. My power had died. Fortunately, when the sun came out later today, it heated the house so I didn't have to worry about the pipes while I went in to the school. At school, everything was controlled chaos with all the little kids outside and the older students upstairs in study halls. Without

power, the school had no water or, more importantly, bathrooms or lunch capabilities, so the kids went home by noon.

It is dark and it is only 5 in the evening. Luckily I do have some candles and water jugs put aside, as well as a woodstove. I'm very thankful that I was able to find a house with a woodstove or I'd be out of luck right now. I'll be getting up every few hours through the night to add more wood to prevent me and my pipes from freezing. It is all an adventure...

EDUCATION GOING BOTH WAYS —
NORTH HAVEN
Keith Eaton

Jan. 1:

You want something done? When it comes to the Internet, sure, I can navigate it, know plenty of web sites and the theoretical aspects of hypertext and what not, but I've never learned html, or the basics of site design, even. Neither has Ian, my student. We want to build a site to describe the school's recently completed boat, RECOVERY, and we'll learn as we go along. Claris Home Page isn't that different than word processing, and Ian is jumping far ahead of me, impatient as I follow every step of the tutorial. In his mind, he's five steps ahead, uploading image files of the boat and prepping an MPEG video for on-line viewing. I want to make sure that we know the toolbar first.

But this is it, education going both ways, student teaching teacher teaching student, in an environment where if you can't accomplish a task for yourself, or can't find someone with whom to collaborate, it most likely will not get done. Besides, chances are that anything you want to accomplish isn't that difficult. People design these things, thus people can figure them out. When my car wasn't starting, D helped me isolate the problem, taking apart the dashboard and removing a squirrel's nest of wires surrounding a jerry-rigged alarm system. Normally, I wouldn't have dared. In lieu of a tow onto the ferry for an "expert" mainland opinion, our troubleshooting made sense; and my, didn't she handle nicely in the snow today!

Island Institute Fellows, 2000–2001

EVERY DAY I LEARN SOMETHING NEW ON **LONG ISLAND**
Dana Leath

Long Island, located about four and half miles off Portland, in Casco Bay, is approximately three miles long and less than a mile wide. About 200 people are year-round residents of the island. Every day, I learn something new about the islanders, the island and the community's history.

I work at the 16-student, two-classroom, K-5 Long Island School. While the regular teacher is out on maternity leave, I am teaching the physical education class twice a week. In addition, I help with reading, math, writing, recess and other school activities as needed. I enjoy being back in elementary school and it is interesting to see how small island schools operate....

The other major component of my fellowship position is sea sampling — measuring and recording data about lobsters. The data is used by the Island Institute and the Maine Department of Marine Resources to help understand and assess the lobster fishery. This is valuable information because lobsters are a critical compo-

nent to the economic vitality of most island communities. After I complete sea sampling training, I'll go out on a lobster boat from Long Island once a week and collect data.

With time, once I know the community better and people here are more familiar with me, I will be able to get involved with the community in other ways. I think the biggest challenge will be trying to start a program or make a difference in a way that is self-sustaining and will continue on after my fellowship position is over. I like it here....

Editor's Note: Of the ten Island Fellows above, Kathleen Reardon is in a master's program at UMO and working for Maine's DMR; Nate Michaud is the Island Institute's Program Director; Dana Leath is a Senior Fellow in Portland coordinating Casco Bay Fellows; Mike Felton is the Institute's Education Outreach Officer; Jessica Stevens is a Community Marine Steward living on Monhegan; Emily Graham, after three years on North Haven, teaches in South Portland; and Keith Eaton is a teacher on North Haven.

CHAPTER FOUR

Fish, Fishermen & Aquaculture

NOTHING BEATS A GOOD FISH STORY. Ted Ames: "I'd heard the story before, but had always taken it with a grain of salt. The idea that a large school of giant cod could have survived in an area that had been heavily fished for hundreds of years seemed to stretch the truth a little bit too far...."

Ames was gathering information about historic spawning grounds when he ran into this particular tale. "William 'Killer' Smith, an old fishing colleague of mine from Jonesport, took me to interview Roger Beal Sr., a retired fisherman," Ames recalls. "Roger (I hoped) would tell us about some of the old cod and haddock spawning grounds that used to exist in the area. Roger finally put the issue to rest. The Machias Bay spawning run of giant cod had actually happened."

Momentarily distracted from his research, Ames wrote this marvelous fish story for the 1996 *Island Journal*.

Mike Brown, a fisherman himself, recalled what it was like to grow up with a fisherman-father and the heritage that experience has left him. The passion in his story is palpable.

Finally, there is marine science as it pertains to fisheries, and the natural tension between scientists and fishermen. *Island Journal* has paid particular attention to the fisherman–scientist interaction over the years because it is so important — to the scientists (who know their biology), to the fishermen (who know their fishing grounds), and to everyone else in a region that depends heavily on the health of its renewable marine resources.

David D. Platt

Dark Harbor Fishermen (detail), N. C. Wyeth, 1943
Above: Peter Ralston

Fishermen's Sons

Mike Brown

If the work called fishing-for-a-living is headed for hell in a scale basket, then it's not the fault of the salt main of Maine where there is still the father-son equation heritage that is alive and nearly well. Oh sure, times change; the cotton has turned to nylon, the cedar to plastic, the make-and-break to turbocharged and real money to script catch-of-the-day. But it seems that in the hearts and heads of many Maine boy fishermen there is the same devotion and respect for their father teachers that were there when the times were harsher, the seas bigger, the boats smaller, the soup a little thinner.

I remember growing up on the coast of Maine as a boy weir fisherman. My dad, the Old Man of my life, weaned me from my mother even before the teachers staked their Three Rs claim to my small body. By the time I entered the regime called school, the Old Man had taught me the ways of fish as they made their mysterious and silent Odyssean journeys to and from lands way beyond my comprehension. It was there, the Old Man said, in places where the icebergs calve and toothed walrus dig clams; and big, hairy white bears lie like pounding cats beside a seal hole in a manless white bay with no name that fish came from.

The Old Man taught patience when only a few fish were reward enough. Sitting there on a cold weir pocket frame on early April mornings with only the wind child, born of land and sea, as a companion to us, the Old Man would teach by silence.

I, the boy fisherman and a student of the times, would have my hands down inside my worn rubber boots as they were cold from the breath of the cross child. And my hand-knit collared sweater would be cinched high with a big blue moon button found and claimed from an abandoned attic sewing basket.

Sanford, 'Dick 'Lunt with great grandson, Nate, Frenchboro

PETER RALSTON (2)

And overall, Christmas-present oilskins now mottled yellow with rutted work lines creased like country roads across a giant field of wheat.

I would cuddle closer to the Old Man and feel his comforting warmth — a presumptuous move of emotion. It seemed the thing to do. And when the wind was cold and gray, sunless emerging light would become unbearable and threaten to defeat my defiant stance, the Old Man would say, "Look, look down there, did you see it?" I would uncoil like a spring and stare with no less intent than a fishhawk into the gray, steely waters of the weir pound.

"There, swimming this way, flashes," the Old Man would say. And I would see them, signs like the scratching of big, wooden matches to light the kitchen kerosene lamp that would begin the new fishing day. I saw them! I saw them!

The Old Man would usually say, in those first run-of-herring April mornings, that there were too few to seine but, son, they were coming, the fish

"By the time I entered the regime called school, the Old Man had taught me the ways of fish."

were surely coming, and tomorrow will bring more to our small cove herring weir of wooden oak stakes and straight ash ribbons and white birch brush. There was always the better tomorrow....

I remember the years that the herring never came in April. We ate a lot of shortbread and pork scraps. The Old Man sure could make shortbread in that black, flat oven pan heated in a Clarion kitchen stove stuffed full of dry alders. And black coffee. I bet I drank more coffee grounds before I was ten than most ashore kids drank milk the rest

of their lives.

When the herring didn't come, the mackerel had to. The Old Man wished them in. Wish the mackerel would hit tonight, he'd say at suppertime. Wish they would too, I'd wish.

Then, one of the days of late June, the mackerel obliged our wishes. There they'd be, swimming around the weir, puffing out their silvery gill plates, looking tuckered and tired from their long swim from way there to way here. Up past that faraway island that looked like a black bowler hat. That island, the horizon of my weir world.

How I loved to catch mackerel! The Old Man and me would unreel the sweep seine from the scow. It reminded me of when the Old Man made special marblecake and folded the batter from the bowl to the baking pan. Like a cattle roundup, we'd herd the mackerel into the weir pocket and then dry the twine and bail the irradiant green striped fish into the dory where they'd pound the cedar planking in tattoos of death. I felt sorry to end their Odyssey.

Ashore, the fish peddlers in old pickup trucks with wooden boxes in their aft pouches, like kangaroos, would be waiting. How I loved fish peddlers. The Old Man and me would lug the fish up the shore flats in peach baskets with stove pot holders stuck in the wire handles so they wouldn't cut our hands. Sometimes the peddlers would laugh and speak in Italian or French and ruffle my hair. The Old Man would laugh and punch me gentle on the shoulder. I'd laugh, too. I thought — maybe I'll be a fish peddler when I grow up. I'd tell the Old Man that later and he'd ask me to say mackerel in Italian. I couldn't.

We grew and shared, the Old Man and me; neither knowing that our lives were being played on similar stages, with like actors, all across the fishing theater in which we all together existed....

One morning, it was late September, and the morning light of that month being cruel had driven me deeper into my hand-quilted cocoon. The Old Man shook me and said it was time to check the weir for the very last time that year. I sure hated to shed my warmth but threw the covers aside and ran down to the kitchen to dress in front of the woodstove that had been brought to cheerful life by father fisherman. Drinking black coffee and eating a toasted biscuit breakfast, the Old Man looked at me sitting there nearly lost in my wrappings of wool clothes and asked me if I thought I'd be a fisherman when I grew up.

If you will teach me, I replied.

1984

JEFF DWORSKY

PETER RALSTON

IT WAS GOOD AND IT WAS ENOUGH

From Amaretto, by Joe Upton

Morning with a hot fire in the stove, listening to the radio at breakfast and watching the day come across the land and the water through frost-laced windows. A walk through the woods to town, past dories and snow-covered lobster traps. Sitting in the one-room general store, nursing coffee and listening to the drone of the old men speaking of fishing stories I had heard before and would hear again and again. Powerful, almost mythical stories of great schools of herring caught or lost. Or stories of the hard winters past, of '36 and '38 when the ocean itself froze out to the islands, eleven miles offshore, and the foolhardy drove their cars across, and all up and down the coast the steamers couldn't land and passengers had to walk out over the saltwater ice far from shore to get aboard.

And when at last we'd slip out, walk back in the cold, snowy dusk, happy to make it to our own house, it was good and it was enough.

1990

93

Less Is More

Cynthia Bourgeault

Three years ago, in a move many would have considered impossible, the 40-some fishermen of Swan's Island, four miles off the southern tip of Mt. Desert, won permission from the State Department of Marine Resources (DMR) to draw a legal boundary around their waters and within its confines to implement Maine's first official trap limit as a six-year pilot project. When it went into effect in 1984, limits were set at 500 traps for a boat fishing alone and 600 for a boat with a sternman. Each year those numbers have been coming down by 50, until they level off in 1988 at the target quotas of 300 and 400. After two years of operation at this level, islanders will have the opportunity to evaluate, and if they so choose, to extend the limit for an additional six years.

"What we're trying to do is to prove you can catch as many lobsters with fewer traps," says Sonny Sprague, 44, Swan's Island's burly, sandy-haired First Selectman, widely considered to be the prime mover behind the island trap law. And indeed, they may just succeed....

Perhaps the most remarkable aspect of this project is the direction from which it originated: not from the top down, from a panel of legislators or experts, but from the bottom up, as a grassroots response to a situation Swan's Islanders knew only too well. Despite a greatly increased fishing effort, the lobster

Little Potato Island, Deer Isle Thoroughfare

PETER RALSTON (2)

catch was slowly but steadily dwindling....

Like so many of our modern conundrums, the source of the problem ultimately lies in technology, which, in Sonny's words, "turned us into over-efficient lobster-catchers." Back when he got his start in the business, it was with "about a hundred traps and an old skiff with a 10-hp outboard. Half the time the engine would break down somewhere out in the bay, and I'd have to row home."

But even in those days, the early '60s, that style of fishing was fast fading into nostalgia as a series of innovations brought a steadily advancing sophistication to the lobster fishery. Hydraulic haulers made it possible for one man to do the work of two. Fiberglass boats and diesel engines doubled the territory that could be covered in a day. Loran and radar took the guesswork out of setting traps and finding them again, even offshore or in "thick o' fog." Most portentous of all were the wire traps, which made their widescale appearance on Swan's Island in the late 1970s.

"With the old wooden traps, you'd have to tend them every day or so or they'd go sour — they wouldn't fish anymore," Sonny explains. "But the wire traps will go on fishing indefinitely, and they don't have to sit on the bank once a year to dry out. With wire traps, it means you can keep your entire string fishing all the time."

This sophistication had a price tag, of course: $60,000 for a fiberglass boat, $1,000 for electronic gear, $30 to $40 apiece for bare wire traps. Add to this potwarp, buoys, baitbags, haulers and associat-ed gear, and suddenly lobstermen were staring at a capital investment of easily $100,000. But with it came enormous leverage as well. Suddenly the age-old natural limit — the number of traps a man could haul in a day — was a limit no longer.

"There were plenty of days when I didn't even make enough to cover my bait and fuel," complains Steve Wheaton, 34, one of the island's hard-driving younger lobstermen — "and that's not even starting to think about my $3,000 boat payment twice a year." It was a vicious circle, and left unchecked, its outcome was only too clear. "Just like in agribusiness," Steve points out. "First the smaller guy is forced out, then the whole industry is pooched."

To win approval for even a limited testing, trap limit proponents on Swan's Island, led by Sonny Sprague and Sheldon Carlson, had to overcome resistance from within the community itself, from neighboring fishing communities and initially even from the DMR, which feared that the Swan's Island proposal might set a precedent for a patch-work series of trap limits along the coast. Perhaps the most formidable challenge was that of negoti-ating a legal boundary within which the trap limit would be operative: a move interpreted by many as giving up traditional fishing grounds — or worse yet, handing it over to state control. It took two years of grueling negotiation — both formal hear-ings with the DMR and informal horsetrading behind the bait barrel — before the fishermen on Swan's Island finally won their way. Even then, it

PETER RALSTON (2)

was by no means a united front....

Perhaps the most significant discovery to emerge out of the Swan's Island project, however, is that a trap limit, even without a limited entry, does act as a significant conservation measure. It restores that imbalance created when technology enabled fishermen to extend their range beyond what they could fish in a single day.

"Sure, it's possible to fish a thousand traps, haul 'em once a week," Sonny Sprague concedes. "But that kind of saturation fishing is incredibly wasteful." And he makes clear that he's not just talking about increased overhead for fuel and bait: lobsters themselves do not fare well when traps are not tended every day. "To show you what I mean, there was one week last fall when town business and the weather kept me in the harbor for almost a week. When I finally got out to haul, nearly half the lobsters in my traps were dead. They'd eaten up the bait and started in on each other. Now imagine what would happen if I had so many traps out that regularly took me a week to get around to them all!"

Stevie Wheaton puts the point even more forcefully: "That kind of fishing is just a waste of fishing grounds — like throwing money away. It's a sin to ruin a natural resource that way."

"There's a difference between just hauling traps and fishing your gear," explains Sput Staples, 29, another younger fisherman, who may already owe his livelihood in part to the trap limit. "If you throw traps out everywhere, something's bound to crawl into a few of 'em. But in earlier days, a good fisherman could take a small string of traps and really make them count for something just by working his gear — knowing the bottom and where the lobsters were likely to crawl to next. It's a matter of instinct and skill — that's what made a good fisherman, not a huge string of traps."

And this past summer, for the first time, most Swan's Island fishermen were seeing enough of a dent put into saturation fishing for those age-old traits of instinct and skill to come back into play.

1987

DRAGGING

Durward: Setting his Trawl

A whole week. Out of
the north all day.

A dry cold. The wind
clean as split oak.

Dark islands, dark as
the march of whitecaps.

Under hills hard on
the shoreline: churches,

settlements, planted
like bones. Out here,

the boat on good marks,
we let the wire out:
the drag plunges and
tugs. First light to

first dark, we tow, dump,
set, tow. Numb to what

cuts our hands, we set,
tow, dump, mend; tow until

dark comes down. We clean
the catch heading in

through dark to the thin
walls of our lives, grown

PETER RALSTON

numb to the wind, numb
to the dark, to all we've
dragged for and taken,
shells returned to
that other dark that
weighs the whole bottom.

Durward: setting his trawl
for haddock, and handlining cod
a halfmile east of Seal Island,

twelve miles offshore in fog.
Then his new engine went out.
A Rockland dragger spotted him,

two days later, drifting drunk
off Mount Desert Rock. He was
down to his last sixpack.
After they towed him back in,
Ordway kept asking him what
—those two days— he'd been thinking.
Nothin. I thought about nothin.
That was all there was to it.
Ord said, Y'must've thought something.

Nope, I thought about nothin.
You know what I thought,
I thought fuckit.

Philip Booth, *1989*

Horatio Crie and the Double-Gauge Law

James M. Acheson

Most of the world's major fisheries are in a state of crisis. The Maine lobster industry is very unusual in that it has come back from the brink of extinction to become one of the most highly productive fisheries in the world. Between World War I and World War II, lobster catches had fallen to five to seven million pounds — so disastrously low that hundreds of lobster fishermen were forced from the industry. By 1994, catches had increased 800 percent to 40 million pounds of lobster, valued at over $100 million. Even more remarkable for experienced observers of fisheries is the fact that despite heavy fishing pressure, the lobster catch has remained relatively stable since 1947. Maine lobster is truly the "comeback kid" of major fisheries.

This remarkable turnaround is due in no small degree to the efforts of Horatio Crie, who led the Maine Sea and Shore Fisheries Commission from 1918 to 1935. It was Crie who was able to get enough support in the industry and the Legislature to pass the "double-gauge law," a controversial measure protecting both juvenile and large breeding-size lobsters; it had been proposed 40 years earlier. Shortly after this law went into effect in 1934, catches began to improve and the upward trend has continued to this day.

Decades later, conservation efforts were enhanced by the passage of other laws, but the double-gauge law marked the turnaround for the lobster industry. It remains the backbone of conservation efforts to this day. No less important, Crie had enough support to begin vigorous enforcement of conser-

ALL PHOTOS COURTESY OF JAMES ACHESON

Hauling by hand circa 1900

vation laws and still keep his job. In an era when virtually all major fisheries in the world are in decline, the significance of Horatio Crie's accomplishments extends far beyond Maine....

In 1918, [Crie] was appointed one of three members of the Sea and Shore Fisheries Commission. One year later, lobster catches declined sharply, and they remained at record low levels until World War II. In 1920, Crie became director of the commission, a job he held until 1931. From 1931 to 1935, he was commissioner of the newly created Department of Sea and Shore Fisheries, at a time when the lobster fishery was truly in desperate condition. Crie was to preside over the state fishery bureaucracy during the worst years the lobster industry had ever experienced.

Fortunately, he was equal to the challenge....

Commissioner Crie was not in favor of lowering the minimum size to nine inches, believing this would not conserve the breeding stock and would ultimately lead to the demise of the industry. He steadfastly favored the double-gauge law, which he saw as a compromise that had something for all factions. He pointed out that the double-gauge would protect small lobsters and conserve the large, prolific lobsters. It would also allow Maine fishermen to catch smaller lobsters, which would enable them to compete more effectively for markets served by Massachusetts and the Canadian provinces. It would also, he argued, keep out 40 percent of the Canadian lobsters that were currently flooding the American market. In short, Crie

Hauling traps with early roller gear

argued, the measure would be good both for conservation and for sales. Crie stated his support for a double-gauge law with a nine-inch overall minimum and 13-inch maximum in all of the coastal newspapers. He also sent an explanation for his position to fishermen and the Maine congressional delegation.

A meeting was held in Augusta in January 1933, to discuss the double-gauge proposal. It had some support among dealers and people from the westernmost counties, but the majority of the people present were against any change in the law. Crie then sent out a questionnaire to all lobstermen; the returned cards revealed that the industry

was badly split on the issue…. In general, the fishermen in the western counties favored the small, nine-inch minimum; men in the eastern counties wanted the existing, ten-and-a-half inch minimum. Many of the dealers were in favor of the double-gauge since they wanted smaller lobsters, and the "jumbo" lobsters found a poor market, anyway….

In December 1933, the Maine Legislature met in special session to deal with a number of issues concerning the deepening economic crisis. One of the industries that claimed its attention was the lobster industry. The double-gauge bill was proposed as the solution to the problems of the industry, and Commissioner Crie was asked to comment

on the merits of this bill on December 11, 1933. He spoke forcefully about the need to preserve the industry by protecting the large reproductive animals. He promised, "If a double-gauge measure is passed…you will see the lobsters continue to increase from year to year and no one will ever have to feel disturbed about the depletion of the lobsters on the Maine Coast so long as a double-gauge measure is enforced." (This proved to be a very astute prediction.)

Even though there was a great deal of uncertainty and disagreement in both the industry and legislature, the views of Crie and the proponents of the double-gauge measure were to prevail. Without much press attention or public debate, the Maine Legislature narrowly passed the bill providing for a 3-1/16-inch minimum and a 4-1/2-inch maximum. Its passage gave Maine the only double-gauge law in the world. It was truly a radical piece of legislation, and remains the foundation of lobster conservation efforts in Maine to this day….

1997

The Great Machias Bay
Cod Run

Ted Ames

This event is supposed to have occurred just outside the harbor of Jonesport, Maine, where large schools of giant codfish were discovered by a dragger at the entrance of Machias Bay as he prepared to go fishing to the "west'ard." The legend goes something like this: After towing his net for a short while just outside Jonesport harbor, the skipper hauled back. To his surprise, his net was filled with giant cod averaging over five feet in length. The net held so many fish that he couldn't hoist them aboard. So he strapped the net to the side of the vessel and towed it back into the dock to let it ground out on the ebb tide. When the tide was low enough, he emptied the net by (depending on who tells the story) either loading trucks or hoisting it back over the side.

And set off a great fishing bonanza.

I'd heard the story before, but had always taken it with a grain of salt. The idea that a large school of giant cod could have survived in an area that had been heavily fished for hundreds of years seemed to stretch the truth a little bit too far. Until last winter, that is, when William "Killer" Smith, an old fishing colleague of mine from Jonesport, took me to interview Roger Beal Sr., a retired fisherman. Roger (I hoped) would tell us about some of the old cod and haddock spawning grounds that used to exist in the area. Roger finally put the issue to rest. The Machias Bay spawning run of giant cod had actually happened.

Roger had been fishing that winter with his father, John Beal, captain of a new sardine carrier-dragger, ROYAL. They had discovered the spawning run, and he shared the adventure with us.

During the winter of 1941–1942, dragger fishermen throughout Maine were having a banner year. The country had just shifted into a wartime economy, and fish prices were sky-high. The Gulf of Maine redfish fishery had just begun, and any boat big enough to tow a net was rushing to convert to dragging.

The 74-foot ROYAL was launched in Thomaston late in 1941. As soon as she was ready, Captain John Beal and his son Roger brought her back to Jonesport. In a short while the new carrier was trans-

SIRI BECKMAN

SIRI BECKMAN

porting herring to the company's Yarmouth factory.

All had gone well until one day when the ROYAL pulled into the wharf, her decks awash with herring, to find the factory "burned flat to the ground." It was a serious matter, for until the factory could be rebuilt, there was no longer a need for the new sardine carrier.

Instead of leaving their brand-new carrier sitting idle during reconstruction, the owners decided to join the redfish bonanza. John and Roger were directed to take the ROYAL to Portland to be rigged over for dragging. The ROYAL could be switched back to carrying herring later on.

By early winter they were towing for redfish.

"I hadn't fished too much before that," Roger recalled. "We were fishing out of Vinalhaven for redfish that winter. There was fairly good fishing around there, oh, about an hour's steam outside Matinicus," said Roger. "But my wife, Buelah, got sick . . . phlebitis in her leg, you know. I had to come back to Jonesport to help take care of her. So we came back and tried to make a living around here until she got better."

In the Jonesport region, the usually plentiful cod and haddock normally left the shore soundings during the fall, backing offshore into deep water until spring. The winter of 1941–1942 was no different. Fish were scarce around home.

"Most every time we went out, we'd make a tow along that bottom just east of Mark Island. We'd start off . . . oh, off about where Libby Island touches the second notch on Cross Island and then run right for the light. Got a handful of fish, you know. Just barely enough so you'd try it again some other day; but not enough to bother setting back."

The ROYAL settled in for the winter. Fishing became reduced to occasional cold, brittle forays into the bay whenever a clear day punctuated the tedious march of storms. Foul days were often used to work on gear. But even gear work has its limits.

"We'd built up a new net on blowy days while we were waiting, and we were getting anxious to go back to the west'ard and go redfishing again."

Eventually, inexorably, winter wore itself thin.

"Buelah was slow getting better, but she was gaining. By the end of March, we figured it wouldn't be much longer. Well, come morning, April seventh, we run out to the tow and hove the gear over. We dragged for about 20 minutes on the marks and hauled back. We had about 10,000 pounds of cod on that tow. But those codfish were all over five feet long! I don't know where they ever came from, because they weren't there the day before. We split the bag and heisted them aboard. No problem with the first tow. Everything worked fine. But we knew that the old net probably wouldn't stand another tow like that. So we run back into the harbor to Charlie's Crick and put on the new net. We kept the old bull rope. [A bull rope is a heavy line with a small loop in each end that goes along the outside of the net from the front to the cod end at the back.] When we rigged the cod end, instead of changing it, we just run the new splitting strap through the eye of the old bull rope and went back out again. As it turned out, we shouldn't have.

"We run off the marks and set back out and towed for about an hour and 20 minutes. This time when we hauled back, you could see the net coming a long way down."

The ROYAL was an eastern-rig; the net and trawl doors were mounted on one side of the vessel. When the captain hauled back, he would lay the boat downwind to help keep the net away from the propeller. This made the vessel drift away from the net while tending the gear and taking the fish aboard. But if you looked down over the side as the net was being pulled up, you were looking right into its mouth.

As Roger watched the net loom into sight, it appeared large and luminous white, and he knew at once it was the reflection of more giant codfish being drawn from the depths toward the boat.

"It looked like a big white ball coming! All you could see was a big white ball! And the whole net was filled from the cod end right to the mouth with those big codfish!"

Finally the trawl doors were hauled into the galluses and the net hung alongside. The Beals worked fast to keep the big catch from swimming and floating back out the mouth of the net.

In an instant Roger had taken the jilson, a line running through a block in the rigging with a small iron hook spliced in one end, and hooked it into the middle of the heavy roller section. As he did, his father took a few turns around the winch head and slowly hoisted the heavy rollers aboard.

Now the giant cod were safely trapped. Now the fishermen were ready to start hoisting fish aboard.

Excitedly, Roger unhooked the jilson from the rollers, untied the bull rope from the net and slipped the jilson hook into its loop. His father hoisted again; hauling the manila bull rope in until it drew tight around the submerged net at the top of the cod end. Slowly, as the splitting strap began to draw tight, the cod end was pinched off from the rest of the net. Excess fish slid forward into the belly.

As the cod end was lifted higher, a bulging bag of fish slowly emerged from under the loose twine and squished heavily against the side of the ROYAL. Finally it swung aboard. Roger reached underneath the massive, swinging bag and grabbed the tail of the pucker string. He yanked hard to untie the knot and jumped aside as several thousand pounds of cod spilled out into the deck checkers.

Quickly retying the cod end knot, Roger pushed the empty bag toward the side and guided it back overboard as it was lowered. Once the net was back in the water, his father steamed the

ROYAL in a circle to wash more fish back into the cod end. When it was filled, he kicked the engine into neutral and hauled the next bag of fish aboard.

"We got two heists aboard before the bull rope parted," Roger recalled. "Then, it let go. When it did, the cod end just hung straight down over the side, underwater. The splitting strap was on the cod end, so it was underwater, too. So we couldn't reach the splitting strap to rig a new one. By then it'd started to breeze up from the sou'west. And you know sou'west is out there . . . . Well, he couldn't do nothing so we strapped the net to the side and towed it all the way from Libby Island into the crick down below the house, here.

"When the tide dropped enough so's we could get at the splitting strap, we hooked in the jilson.

"They were some heavy, too," Roger went on. "I was winchman and when I took a strain on the bag, you could see it was gonna be too big to heist. We had a 60-horse Kermath on board for a donkey engine and that thing was turned right up. When he said, 'Put an extra turn on the winch head and come back on her,' I did.

"Well, that hook on the end of the jilson straightened out and shot up into the rigging like a cannon! It's a wonder it didn't kill somebody. After that, we took them aboard in smaller heists.

"We had to split them three more times before we finally got them aboard. We'd towed them in so hard they were packed solid. Best as I can figure, there was about 30,000 pounds in that tow."

The very next day on the first tow, they hauled back yet another huge net of fish. Once again they broke the bull rope; again they had to tow the net back to the harbor, this time towing it to Layton's Wharf.

Total time fished: about three hours. Total pounds landed: between 80 and 90 thousand. This was no stray school of fish.

When they returned to Libby Island Sound, two draggers from Portland had arrived. It hadn't taken long for the word to get out. Soon a formidable fleet gathered to fish on the giant cod.

The spawn fish, the draggers soon discovered, were not only filling Machias Bay, but were also abundant all the way from below Moose Peak Light in the west, to the south of Cross Island.

The end was predictable. The fish lasted through the spring of 1942. The following spring, even more vessels arrived, but caught fewer fish. By the third year, the spring run of giant Machias Bay codfish was broken. The bonanza was over....

1996

Needing Each Other

James M. Acheson

I first began gathering data from lobster fishermen along the central Maine coast in the late 1960s. Compared to the Purepeche communities in highland Mexico, where I had previously worked, the field work was easy. Everyone spoke English; the roads were paved; the water was drinkable; there were police in every town; no one had a gun. The anthropological work I was doing was generally non-threatening, and I had no trouble getting interviews.

I became fascinated with the territorial system of the lobster industry and the implications of this system for conservation. I was only vaguely aware of the rift that existed between the fishing community and scientists who work for federal and state fisheries management agencies.

In the mid 1970s I worked for the National Marine Fisheries Service in Washington, D.C., where I came in contact with many of their scientists. I saw things from their point of view. In particular, I was horrified to learn of stock assessments coming out of the agency's Woods Hole laboratory predicting a lobster "crash" due to excess effort and inadequate attention to conservation.

Several different fishermen, meanwhile, told me they thought that the fishery was sound and that current legislation was working well. Still, I was a believer in science, and so I spread the word of the impending disaster to all who would listen.

When several years went by without the predicted disaster having materialized, I began to wonder. The 1990s have seen my skepticism grow, as lobster catches have achieved levels not seen for a century,

PETER RALSTON

Stonington fish house

JEFF DWORSKY

PETER RALSTON

while the best biologists have continued to predict calamity.

I am now convinced, along with many fishermen, that some of the predictions of scientists are not based on the best information — and that the predictions of fishermen are buttressed by far more than folklore.

Fishermen and scientists, in their worst moments, tend to vilify each other, but I am convinced that the fisheries problem cannot be reduced to a morality play with one side representing evil and falsehood and the other virtue and truth. On one level, these two groups have very different interests: fishermen need to catch fish, while fisheries managers are employed by regulatory agencies whose job it is to enforce laws designed to curb fishing effort.

There's more to it, of course. Fishermen and biologists have very different views of the ocean, marine resources and the motivations of the humans who exploit them. They come from different cultures; they have different experiences, live in different occupational organizations. Most

important, they interact largely among themselves. Over the course of time, out of the same ocean, they have constructed two different worlds.

Robin Alden, former commissioner of the Maine Department of Marine Resources, made much the same point in a 1996 speech at the University of Maine when she said fishermen and scientists are two groups of people with very high respect for their own knowledge, and little understanding of what the other knows. These attitudes have been given a good deal of intellectual respectability by the theory of common property resources — the notion that resources not under private ownership (parks, rivers, oceans, air, for example) may be used by anyone, but are subjected to escalating abuse, because those who use them are strongly motivated to over-exploit them. Why should a skipper conserve a school of herring when it will just be taken by the next person who encounters it? Only rules imposed by government can avert disaster.

Fishermen will often describe fisheries as going in "natural cycles" with fish stocks rising and falling, suddenly and unpredictably. Some of them will say that the factors controlling the size of fish stocks are so complex that humans have difficulty understanding what is happening. The idea that humans can control the size of fish stocks by controlling a single variable, such as fishing effort, seems ludicrous to many fishermen, to say the least.

Typically, people who live from the sea know a great deal about the life cycles of the species they exploit, including spawning behavior, nursery grounds, migration routes, growth rates and predation. The most important variables, in their view, are natural phenomena such as changes in water temperature or predation by other fish. Human behavior can affect these life-cycle processes, but usually in a less important way. (This point of view coincides with that of some academic biologists trained in ecology.)

Fishermen typically have very fine-grained knowledge of what influences the stocks of fish in the areas where they fish. They know that certain life-cycle processes of a species may exist over its entire range; other species may be strongly affected by variables in a small area. It follows that many rules need to be highly localized in nature.

Fishermen are convinced that stocks vary because of changes in factors affecting the life cycle. They regard rules designed to maintain life-cycle processes as effective, and they are much more likely to obey such rules since they believe it's in their own interest to do so.

A University of Maine biologist and I recently looked at the explanations fisheries biologists and fishermen have given for the marked changes in lobster catches seen in the 20th century. (In the lobster "bust" of the 1920s and 1930s, lobster catches were one-eighth of what they have been in the "boom" years of the 1990s.)

The biologists tended to explain changes in lobster catches in terms of fishing effort (e.g., number of traps) and water temperature. The fishermen explained them in terms of types of fishing practices and environmental factors. Fishermen's favorite explanation for the "bust" was massive violation of the conservation laws and the so-called "poverty gauge," the large legal minimum size gauge which made it illegal to keep most of the lobsters caught.

The most popular explanation for the boom of the 1990s is the absence of predation by groundfish, which are currently at low stock levels. Other explanations include the oversize law, which makes it illegal to take lobsters over five inches on the carapace, and the v-notch law. (Fishermen may not take female lobsters with eggs. Voluntarily, they mark "egged" females with a notch in the tail. Such v-notched lobsters may never be taken.)

These laws, fishermen believe, are very effective in protecting the broodstock, which, in their view, is the secret to a healthy fishery.

But neither side in this debate had a premium on truth. The data buttress some of the ideas of both fishermen and scientists about the causes of the changes in lobster stock sizes; other explanations are clearly fallacious. This situation, in the Maine lobster industry, does little to build bridges of confidence between fishermen and biologists.

Yet there are hopeful signs. Among academic scientists and some in the management community, there is a resurgence of interest in ecology and studies of the life cycles of fish. Scientists of this persuasion will find it much easier to communicate with the fishing community, since both groups have a more similar view of what controls fish stocks and management goals.

Virtually all traditional tribal and peasant societies in which people exploit marine resources have rules to conserve those resources. To be sure, many traditional conservation rules have eroded under the influences of modernization. But no maritime societies, even those with no scientists, permit the unrestrained destruction of the marine resources on which their livelihood depends.

It is much easier to generate effective legislation if fishermen and scientists are inclined to cooperate. In this regard, the history of Maine lobster legislation is instructive. Most lobster legislation was initially suggested by scientists concerned with conservation, but passed into law only after powerful factions in the industry came to the conclusion these rules would benefit them economically. The minimum size measure and the prohibition on taking egged lobsters came about in this way.

In 1979 the escape vent law came into being by a completely different process. It passed through the legislature with unusual speed due to the strong support of the Maine Lobstermen's Association and the Department of Marine Resources. Several months of discussion involving fishermen, scientists and legislators led to a consensus that allowing undersized lobsters to escape from traps would reduce the mortality of juvenile lobsters. Everyone agreed it would be good for both conservation and efficiency.

The rapid passage of the escape vent law suggests what might be possible if the goals of fishermen coincided with those of scientists. When the two groups are able to communicate and broaden their perspectives, the opportunity for effective regulations grows accordingly.

1998

Running Together

K. J. Vaux

Inside the Black Prince weir, nine men with their skiffs tied together sit captive to a story being told by the oldest among them. Animating his story with hand motions and island lingo, the senior man's is the only audible voice. As he delivers the punch line, a burst of ho-hos and ha-has bounces off the cliff directly behind the weir. Cigarette butts get tossed into the water; three men move their skiffs and hop into the boat closest to the net inside this giant herring trap. Together they pull hard on the net, the tiny skiff heaving and rocking as the net slowly surfaces, encircling a foaming pool of thrashing fish. Bringing their skiffs around close, the rest of the men dip their nets into the shrinking pool, pulling out the wiggling silver fish they will use for lobster bait. They dump them into bright green plastic baskets. As the net gets pulled higher and higher, the sound of the herring splashing is rivaled only by the seal-scarer from a nearby salmon farm.

Less than a mile away, salmon leap out of the water, mouths open, as young men spray pellet feed into their underwater cages. Since its introduction to Deer Island nearly 20 years ago, aquaculture has presented this depressed New Brunswick community, just off Eastport, Maine, with a multibillion-dollar industry. But with the promise of economic prosperity came a threat to a traditional lifestyle dominated by herring and lobster fishing for over 150 years. Aquaculture has forced many to take sides, and islanders continue to face the challenge of change as they struggle to maintain the fabric of their community.

The sun rises orange over Campobello Island, starting a day begun hours earlier by Dale Mitchell. The sunrise tips and disappears as the FAMILY PROVIDER, his boat, heaves in the swells. Two small trees he has cut off his property to use for repairing the weir trail behind. With Mitchell is his brother-in-law, Gary, who co-owns the Abnaki weir. This morning they will start preparing it for summer, the season for catching herring.

PETER RALSTON

BILL CURTSINGER

Penned salmon are fed three times daily, sometimes four.

HEATHER HAY (2)

Dale Mitchell has been working these waters for most of his 41 years. "It's all I ever did," he says. "It didn't occur to me to be a noble thing. But my father bred it right in me. And I love to do it. I figure I'm successful at it. It's all I ever knew. It's me."

Before building the weir in 1986, Mitchell spent four years watching the herring move around the tiny islet where it is located. Knowledge of an area is imperative: A fisherman must follow the routes herring travel in order to build a weir that will catch them on rising and falling tides.

Mitchell points out an eagle's nest. "That one's been there about 10 years now," he says. "They get bigger every year. Some eagles' nests have been around as long as my father can remember."

Weirs may be owned by as many as nine or ten people, often within extended families. "It just kind of works out that way," he explains. This tradition of joint ownership has kept the community tight over the years, as the work of one becomes the work of many. "Deer Island is a place where, if people see that you'll work, they'll make sure you

get by."

Behind each weir is the steadfast tradition of its name — for location (Grass Point), for the men who built it (The Bachelor), for historical people or tribes (Abnaki) or events of the year they were built (the Jubilee weir, constructed in the 50th year of Queen Victoria's reign). The Zig Zag weir was named for its shape; the Black Prince weir refers to the color of the water when the weir is thick with trapped fish....

At 7 a.m., Allison Pendleton motors through the dense white fog to his salmon farm a few kilometers from Lord's Cove. Allison has been on the water since he was 15 years old. "I left school one day, and went dragging scallops the next," he says with a smile, recalling the distracting view of harbor traffic from his school.

Allison farms 80,000 salmon in 32 floating cages. Each cage is supported by circular, hollow plastic tubes. Beneath them, nylon nets drop 36 feet, containing his salmon in a cone-shaped area where they are fed three times daily, sometimes four. Today his crew will harvest about 2,000 salmon. Although Pendleton was one of the first Deer Islanders to enter aquaculture 13 years ago, he was originally a traditional fishermen, tending weirs and trapping lobster for nearly 40 years. But like many in traditional fisheries, falling prices and diminishing stocks of herring made it harder and harder to earn a living. "Herring got so bad that we was going seven, eight years, and every year going in debt. I just couldn't go through another one. I woulda been completely broke," he recalls.

Recently introduced to the area, aquaculture offered an alternative. "Most of [the farms], like ours, we had a weir and it wasn't catching any herring so we sawed it down and put an aquaculture site in there." While it was a business with its share of risks, the profit margins looked promising, and his family was willing to go into the business with him.

Aquaculture presented Pendleton with the continued freedom of being his own boss, managing his own business and working on the water. It provided a steady income, without the ups and downs of traditional fishing. "Aquaculture has been good to us. We never get rich, but we make a good living, something we never had in fishing." While the hours are long, and the work is hard, "you know what you're doing. And you're free to do what you want."

Pulling up to the cage he will harvest, his crew brings the salmon to the surface by pulling the seagrass-covered net almost entirely to the surface. Using a mechanized dip net, they scoop a few dozen fish out from the pool and dump them into a holding tank on deck. Two men grab fish and lay them flat, while a third makes a quick cut under the gill and shoves them towards a container of ice and sea water. Some glide listlessly, others kick and flip frantically. Once in the tanks, scarlet plumes rise to the top as they thrash and pump their own blood into the salty ice water. This method not only keeps them alive longer, but leaves less blood for workers to clean at the processing plant.

Cruising home past a dilapidated weir and the salmon farm directly next to it, Allison reflects on the advent of aquaculture on Deer Island. "It's not as bad now as when we started. When we first started it was hellish. People wouldn't speak to you. They wouldn't go to church. Wouldn't associate with you at all. But most of that's gone now because most people on the island are involved in aquaculture in one way or another."

Earl Carpenter, owner and manager of Deer Island Salmon, felt this tension when he first came to the island to establish his business. "In 1985 when we first came here, there was very strong resistance to doing anything differently. The people that were still here, quite rightly, had the notion that it would change their lives if something new came along that was dominant; and change is a hard thing to accept when you don't know what the outcome is going to be." Yet by stressing the importance of community, Carpenter feels he made important inroads among skeptics. "We had a very strong emphasis in the early years of putting money back into the community where we could," he says. "And the people sensed that here, and they went with that. They said it made sense."

What initially made sense to some islanders was the jobs aquaculture promised. The sardine industry declined during the 1980s, and in 1993 the sardine factory in Fairhaven closed, leaving nearly 20 percent of the island out of work. Aquaculture, meanwhile, provided jobs for islanders who otherwise would have had to aban-

Dale Mitchell

don their homes, communities and communal history in search of work. In addition, the salmon industry provided a sense of future for young islanders….

His processing plant is filled with both those left out of work by the collapse of herring, as well as those new to the workforce. Dressed in long yellow rubber smocks and clear blue aprons, eight women stand along a sterile metal trough, each working on a specific task of gutting and cleaning salmon. Beyond them, four young workers label, package, wrap and palletize salmon for the delivery trucks waiting outside. Upstairs in the office, three young women and two men manage the business, contacting buyers and distributors from New York to Montreal.

For those who once packed sardines, aquaculture has meant not just employment, but better wages and far better working conditions. "I used to pack fish in Fairhaven," says Grace Regression, 58. "You stand there eight hours a day and you don't lift your head because the more you pack, the more you make. And that's slave labor, I'll tell you. And they come out of there five years later and their hands all twisted [with] arthritis. This is more relaxed. I get a chance to step outside for a puff," she continues, sharing a cigarette with her co-workers as they stand outside on one of their hourly breaks. Also working at this plant are young adults, too young to have packed fish in Fairhaven, but old enough to have looked for work off the island were it not for the jobs salmon has provided.

"Without salmon here, there wouldn't be any young people left on this island," says Chad Stuart, 26, who worked for five years on a salmon farm.

"People [are] getting all upset about the salmon sites being here, but if you want the people here too, that's how it'll be," Stuart says. "We could be doing the same thing on the mainland as we're doing here. Where would that leave the island? Even if you went somewhere else, where do you start, Burger King?"...

Aquaculture brought the jobs that keep islanders, young and old, at home, but those jobs have also brought outsiders. Many come from nearby mainland communities to work. In particular, Deer Island has a growing population of Newfoundlanders — left jobless by the cod moratorium, closing paper mills and dwindling employment opportunities back home....

Many native islanders are uncomfortable with the growing population of people from away. "I don't know anybody on this island anymore," says an older resident of Leonardville. "They've all come for jobs, but when aquaculture leaves, this island's going to be in as bad shape as before. And we're gonna be the ones left with it. It's our home. Aquaculture was set up for the people of the island for when they had a bad year in the sardine business. Every year wasn't a good year. You knew that when you built weirs. This was the way of life."

The "way of life" for generations of island fishermen has been defined by the work ethic of traditional fisheries, stemming from the self-dependency, ingenuity and independence neces-sary for the wide variety of work traditional fishermen must do. "A fishermen can do accounting, can do mechanical work, can do carpenter work. He can rig twine. Anything he'd need to do, he can do it. He can educate himself to do it if he needed to do it," Dale Mitchell boasts. "The best part of fishing is doing good. I'd rather catch 200 pounds of lobster at $5 a pound than 100 pounds of lobster at $10 a pound. When you're through, you've got more sense of fulfillment, more satisfaction to see that boat loaded. You can say 'Boy, you had a good haul!' Aquaculture brought a new work ethic to the island. "Every other Thursday that check is waiting for me," says Bradley Hurley, who works on one of the salmon sites.

Traditional fishermen have been reliant on a precarious, often diminishing, natural resource. Aquaculture, on the other hand, is internally dependent and industry-driven. The difference between farmers and fishers has sparked conflict on the island as many islanders, weary of an uncertain financial future, were attracted to the stability they saw in aquaculture....

While these hourly wage jobs may provide more stability than traditional fishing, Dale Mitchell worries about the effects on island life and community. "To give someone an incentive to share in the profit, I think it makes a difference. People are much more up front, they keep things up, and they keep things going," he says. "But [in aquaculture] they don't see that the small things they do help the profit at the end of the year, because they don't share in the profit. It's just a job, and it's a boring job, so they're not gonna put more in than they got to. It will become just another place where people work by the hour. I think that ruins a community, because nobody has any pride in what they do. It's just a job. It's not a way of life. The whole way of life is being lost, because people have lost their sense of what it takes to be successful...."

Despite the fractions and deep disagreements that aquaculture has incited on the island, it has coexisted with traditional fishing for nearly 20 years. In that time, islanders have become more comfortable with the industry, and the change it has brought. Even the staunchest traditional fishermen have come to accept that aquaculture is on the island to stay, and that a majority of islanders have benefited, at least economically, from it. Unlike Campobello and Grand Manan, which have sizable tourist industries, Deer Island has retained its identity as a working island. And as aquaculture and traditional fishing innovate and advance, so will the island and its people. Aboard the FAMILY PROVIDER, Dale Mitchell scans a small screen that flashes information from his Global Positioning System. "I didn't want to get it, but

HEATHER HAY

Inside the Black Prince weir, fishermen prepare to pull the net.

you gotta keep up with the times," he says. Dale and his wife often spend hours on the Internet, gleaning all the information they can find to help him keep up with the elements, other fishermen and the salmon industry. Recently introduced to salmon farms on the island are automatic feeders, which change the nature and numbers of work on marine sites. And as fish health and management have come under closer government scrutiny, the battles fought by both industries stand to become increasingly political. Changes will push both industries to innovate and evolve. And in the process, islanders too will also be shaping their identity, one as much born of the past as created by the present.

As the sun sinks behind the mainland, an opaque moon rises on the islands. The soft orange light that marked the dawn falls low on the water, and a boat blades across the reflecting surface

where Abnaki legend claims the god Glooscap turned a pack of wolves chasing three deer and a moose into islands. From the air, the moose (Eastport's Moose Island) leads the three deer (Grand Manan, Campobello and Deer Island) as the Wolves follow in close pursuit, the islets among them dirt kicked up by the deer in their dash for survival. Like their namesake, Deer Islanders continue to keep one step ahead of the forces that threaten their survival. And although initially with trepidation, they have run together with the challenge and opportunity of change within the waters of the world's largest tides.

1999

PETER RALSTON (2)

CHAPTER FIVE

Working Waterfronts

"WORKING WATERFRONT" IS A PHRASE so alliterative and descriptive that in Maine alone, it has been the theme of a statewide referendum, the topic of various coastal studies, and even the name of a newspaper of wide circulation. Places that deserve to be called "working" waterfronts are surprisingly rare in Maine — less than 25 miles of the coast qualifies — and outside forces are shrinking that number. From the start, *Island Journal* has stressed the need to understand Maine's working waterfronts and protect them from inappropriate change. Stories reflecting this interest have appeared in nearly every issue.

The late George Putz understood the meaning of a working waterfront better than most. "Along the waterfront people think differently," he writes. "The very notions of what time is and tides are, separate the mariner from the lubber."

These distinctions extend out to sea where mariners work, of course, as Jeanne Rollins suggests in her account of the pilots who board incoming vessels for the trip up the Penobscot, and they're never clearer than they are in the mind of a summer sailor who finds himself in trouble — as Jan Adkins did in a small harbor full of locally grounded fishermen. Boatbuilders, waterfront kids, the sons of fishermen, those who collect and celebrate the histories of working waterfronts — all have graced *Island Journal* over the years because they illuminate this curious land–sea interface, where the clash and the melding of different cultures forms a unique slice of coastal and island life.

David D. Platt

A Singular Community

George Putz

At the outset, a working waterfront is all plain and clear. There is a harbor with at least some protection from the weather and sea, around which is based a small township. The infrastructure — wharves, piers, docks, ramps, floats, and breakwaters — conjoin to serve marine traffic, boats, and marine trades. The latter may include fish and cargo handling, fuel and bait facilities, seafood processing, chandleries, and marine construction and repair facilities. They are dynamic places, busy and interesting. Differences stand out in both boats and men because of their specialized roles. In all respects it is a singular community, thought of and acting quite independently of other aspects of the town.

Along the waterfront participants in the culture have separate expectations, rules, and language. Mari-time usefulness, marine competence, and nautical canniness override other considerations for citizenship. It is an environment based on laws of the sea, not terrestrial governance or bourgeois piety or goodness.

Dependability, stamina, and skill are more useful on and about the water. So long as they pull their load, reprobates are as fully enfranchised as any other sort of character.

Working waterfronts are physically stressed places. Weather, salt water, marine organisms, and ice continuously chew away at structures that are expensive to build and difficult to maintain. Some level of seediness comes over them, compounding a suspicious notoriety. This in turn is enhanced by the labor- and materials-intensive nature of marine work that often assumes an unkempt quality — tubs, barrels, crates, and traps piled about, gear and engine parts here and there, repairable and discarded equipment lurking over a ground patina of scrapped small parts, rope and wood, with or without a binding sauce of spilt petroleum, seaweed and creature shells, paint chips, and a million odd fastenings, screws, nails, and broken effluvia. What is not seen by the visitor is that everything is in its place, and that mariners in fact are usually as fastidious as an Edwardian dame. Waterfront people are not there for the view (although

Steve Gray, Stonington

PETER RALSTON

JEFF DWORSKY

they like it), piers are repaired before they fail, and the apparent mess has its own logic and order.

Most marine enterprises are built on information — where the fish are or likely will be, where the best bait is available at the best price, what is the market, who is being most successful and how, where can a rare part be found or made, how can something be fixed before the next tide? Waterfronts are built for work, but sociability is inherent. Kibitzers are expected and usually encouraged, so long as they take as well as they give and are sympathetic. Wharf talk is equal parts technology transfer, secret sharing, bravado, commiseration, and guy-talk.

Since everything around the water is in a process of constant disintegration, the waterfront is a place for mandatory creation, generation, renewal, and repair. Mariners live in the midst of their own industrial archaeology, their middens building like snow drifts.

Along the waterfront people think differently. It begins, of course, with the verities of time and tide and the other stuff of maritime homily, but these differences run deep and are not mere sentimental trifles. The very notions of what time is and tides are, separate the mariner from the lubber and the waterfront community from inland communities, even though the latter may begin but a hundred yards back from the shoreline.

Maritime people are most at home along the waterfront. Non-marine parents warn their children not to go there.

A mariner's sense of time is not like that encouraged by other vocations. It goes way beyond not having a time clock overseen by social institutions, for time itself differs. Hours and minutes have nothing to do with it, and even day and night do not impress as time periods. Rather, they are photometric conditions, when the presence or absence of light affect how fish behave.

Purse seining, for example, is conducted in the dark, not at night. Even the idea of "a moment" is metamorphosed by nautical regimen, as sleep is replaced by "kinks," and hourly intervals by tricks

KOSTI RUOHOMAA

Wharf talk is equal parts technology transfer, secret sharing, bravado, commiseration, and guy-talk.

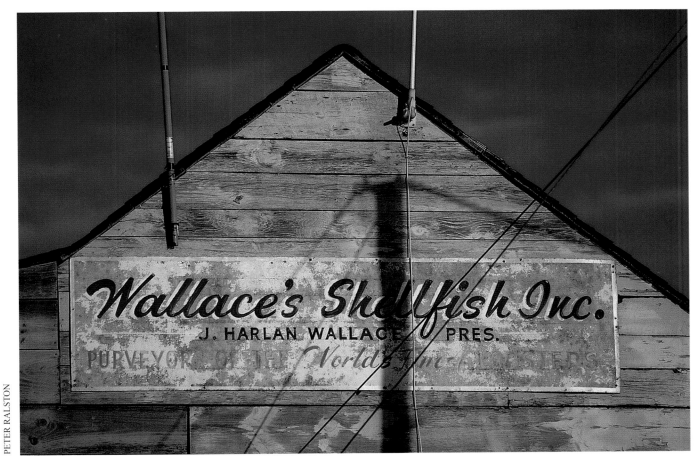

PETER RALSTON

There are no contracts, no regular paychecks, and no employers.

at the wheel, deck watches, sets, tows, haulbacks, and lumping sessions — long stints of picking fish, cleaning, and icing down the catch. Deepwater fishermen often come to ignore time altogether.

This sense of time leads to different views of the world and a disparate attitude about a person's place in history. Terrestrial folks tend to view the universe in terms of circles — down to life cycles, seasonal rounds, and periodic paychecks of an expected amount. They are employed during work weeks separated by weekends, when one's time is one's own. On the other hand, while mariners have their cycles of tide and seasons of different target species, gear rigs, and periods ashore, days, weeks, and months are irrelevant to them. Their cosmology is more lineal, so they emphasize Fate rather than the Destiny celebrated by their lubberly neighbors.

There are no contracts, no regular paychecks, and no employers. Crew and sternmen are shareholding partners. No catch, no money. It is no wonder that such a philosophy, when confronted by constant battle amidst the elements and a product that either sells or smells, has a personal disdain for paperwork and distrust of people who presume to work by creating and shuffling paper. Quite literally they are in a different universe.

In the boatshop fairing a plank, in the bilges dealing with a recalcitrant ignition, on the dock weighing up a trip of fish, at the washboard shocking-out scallops, everywhere about this lineal neighborhood there is a pervading impatience with ideas about ideas (like this essay), and any scientific thought based on presumed reality or on remote authority. To paraphrase President Truman, the general feeling is, "I'm from the waterfront; show me!"

...People who have not done it cannot conceive of the hassles involved in commercial fishing, and people who are involved have a 10,000-year-old divine right to exercise instinctive contempt for the terrestrially comfortable....

No matter how harassed by terrestrial institutions, mariners are hunters and gatherers, and so heroes. Everyone stars in his own drama, is protagonist to his own story, and all heroes strive to meet the expectations of the role, no matter how simple the plot or how humble his place in it. Just to be in the midst, tolerated if not included, surpasses ordinary ambition.

While others are trapped by insane visions of Destiny, on the waterfront men regard their Fate, and share it.

1987

Monhegan Night Watch

Jeanne Rollins

What am I doing at the beach on Monhegan at 10 p.m. in January with freezing temperatures and 48-knot northeast winds? The cloud cover is an eerie gray and the occasional lighthouse beam lights up the angry seas inside the breakwater. I question how badly I want to experience pilot work in the winter.

Capt. Sherm Stanley and his son Shermie, both of medium height with sturdy and trim builds, appear out of the night. Of old families from Monhegan Island and along the Maine coast, they still make their living from the sea. Using his lobsterboat, either one alone usually takes the Pen Bay pilot out to meet a ship that needs to be guided up Penobscot Bay to Searsport or Bucksport. But tonight the weather looks bad and the forecast calls for worse as the night wears on.

With the help of pilot Gil Hall, a veteran of two decades of this ofttimes hazardous work, the four of us carry the skiff across the beach to the rolling sea. It is going to take a real trick to fit all of us into the little boat and go out through the breakers without being swamped. Sherm quickly pushes the skiff between the black waves, and, with his usual politeness aside, he hollers for me to jump in. In the dark I watch the outline of Sherm's oars in the beam of the lighthouse. It is beginning to snow and the tiny flakes sting my cheeks, hinting that this is just the harbinger of an approaching storm. I listen to the Manana foghorn, already a muffled blast absorbed by the snow. Heavy seas bear down on us through the harbor, causing the lobsterboats to tug on their mooring chains.

Alongside the pilot boat PHALAROPE at last, I scoot onto the deck and into a corner next to Gil

PETER RALSTON

124

where I'll be most out of the way. Sherm and Shermie immediately busy themselves getting the boat ready to leave. Sherm starts the engine and checks all his instrument panels as Shermie walks the skiff to the bow where he can drop the mooring chain. Sherm heads out the mouth of the harbor while I press my nose to the windshield and watch the storm envelop the island.

Gil turns on the VHF radio to listen for the expected tanker, the SPRAGUE CAPELLA. This ship, hailing from Liberia, is carrying 40,000 barrels of oil to Bucksport. Gil calls the ship: "SPRAGUE CAPELLA, SPRAGUE CAPELLA, this is the Monhegan Pilot boat, WZW-5628, standing by." His voice is calm, and anyone listening to the conversation could picture the man sitting at home by a warm stove. "This is the SPRAGUE CAPELLA to the Mon-he-gan pilot boat," comes the reply in broken English. "We have been de-layed, our new ETA is 23:30." I groan to myself. It will be an hour's wait on the building seas.

A boat at night is a different creature. To avoid ruining our night vision, the only lights on in the cabin include the soft red glow of the compass and instrument panels, and the orb of green from the radar scope. I strain my eyes to see shapes and figures, but it's useless. We idle about a mile behind Manana Island while Sherm keeps the boat into the seas and maintains headway to keep from drifting off course. My first challenge is to keep my balance at night. While in the daytime I can handle rough seas with little effort, in the dark I stagger around like a drunk. Other senses take over where sight leaves off. I smell the odor of herring pickle, feel the pitching of the boat, hear the rumbling of the engine and the howling wind. Despite the cold winds, sweat breaks out on my brow and seasickness becomes my chief concern.

For an hour we wait until the now-familiar voice comes over our VHF radio. We switch back to channel 10 to hear the latest progress report. "We are now passing the 14 M whistle buoy." "Roger, SPRAGUE CAPELLA, we will do a port boarding. Please maintain a speed of six knots." The port boarding will allow the 675-foot tanker to create a relatively calm lee, and maintaining a slow speed will give the skippers some control in the heavy seas.

As we near the ship, a glow appears through the snow as if the whole horizon were beginning to lighten. Soon we can see the tanker framed by the blizzard like a ship in a bottle. The sea levels out considerably as we drop toward the stern of our moving target. I can see a seaman amidships on the port side lowering a rope ladder over the rail.

Shermie climbs up onto the washboard and skates forward through the slush. He fastens a rubber bumper over the starboard side. The deck lights of the tanker are enough to light up our cabin. With the stern bumper also in place, Sherm speeds up and closes on the tanker. The water boils between his thin hull and the wall of steel a few feet away, and I can feel the tremendous suction caused by the ship's propeller. Our engine echoes loudly off her hull.

We are alongside the Jacob's ladder, and Sherm slows the engine to keep pace with the tanker. With hulls only about a yard apart, Gil climbs onto the washboard to wait. Except for his bright orange flotation jacket, he could be patiently waiting for the next train. With the vessels barely a foot apart, I expect Gil to jump the ladder, but instead he hesitates to study the motion of the boats. Then, when the SPRAGUE CAPELLA rolls into a trough and starts ascending the next sea, Gil steps across the ladder and quickly climbs out of reach of the PHALAROPE. If the vessels were both descending, he'd risk being crushed between the hulls when they rolled together.

Sherm revs the engine full ahead to break out from the suction of the tanker. We pull away as if in slow motion while Shermie again climbs forward to take the bumper. Out of the ship's suction, the engine is slowed and we draw back toward the stern of the ship. Sherm warns me to brace myself for rough going as we come around the stern.

Almost simultaneously the SPRAGUE CAPELLA switches off her deck lights and we are again in total darkness. The wind is now gusting to 60 knots and the seas continue to build. If the temperature drops a few degrees, we'll be in danger of icing up. Finally, the lights of Manana Coast Guard Station pierce the snowstorm. We round Casket Rock without a glimpse of it and head up the harbor. For the first time since leaving, we are welcomed by the lighthouse beam. Except for the ever-present kerosene lamp in Rita's window, the rest of the harbor is dark as Sherm guides the PHALAROPE to her mooring.

I can feel my tension lift as the deck lights come on and my straining eyes relax to the contrasts of dark and light. Skeletons of redfish litter the slush-covered deck. Sherm sets to work scraping them into a shovel.

This is the end of the trip for me, but for Gil Hall it is only the beginning. For over 20 years he has been piloting one or two ships a week using either Monhegan or Matinicus as a base. Along with six other Pen Bay pilots, he has navigated ships and tankers to ports from Rockland to Eastport in all kinds of weather. Delivering pilots to and from these ships is one more way in which offshore islanders serve coastal Maine.

1987

Boats and Hoops

Sandra Dinsmore

"If you kick over a rock, you'll find a boatbuilder on Beals Island," goes an old saying. The boatbuilder adage applies to lobsterboat racers and basketball players as well. Beals Islanders have played on many state high school basketball championship teams, and while fiberglass has eliminated most of Beals's boatbuilding shops, champion lobsterboat racers and basketball players remain the norm for this small downeast community.

Beals, itself, is made up of two islands: Beals Island is about a mile long; Great Wass Island, about seven miles long, connects to Beals by a causeway. The majority of dwellings are concentrated on Beals, and concentrated is the right word: the houses at Mack Point, homemade for the most part, are jumbled together so closely that if you spit out the window, you might hit your neighbor. Property lines appear nonexistent, though everybody knows who owns what, or they don't care. Three active churches foster community closeness of another kind. In winter, piers loaded with lobster traps line the shore on either side of the Jonesport-Beals Bridge, announcing the islanders' main occupation, as do seasonally boat-choked coves.

That 600 people can produce so much success seems due to familial closeness, respect for each other, lives of hard work, stubbornness, determination and competitiveness; all of which breeds confidence. Most Beals Islanders are related, descended from one or two common ancestors. Manwarren (originally spelled Manwaring) Beal, whose family arrived in America in 1621, settled on what was then called Little Wass Island about 1765, changing its name to his; Captain John Alley arrived in 1770. Beals and Alleys still predominate, but other families have added diversity to the island.

Basketball player Sandi Carver, 23, grew up in a gym along with teammate Jan Beal, who's now playing basketball at the University of New Hampshire. After playing high school basketball, Sandi played at the University of Maine for four years and loved the support she got from townspeople back home. In Orono, she feels, parents come to games, but at home, she says, "It's not just your family, it's all of your family and all of your friends and everyone in the two towns; everyone comes out to watch

STEVE CURTIS

the games." She thinks it's because everyone knows and cares about everyone else and, because there are no movie theaters or shopping malls, it's entertainment. Beals Islanders get right into it, yelling suggestions and comments, riding the referees, loving every second.

It's the way they do everything — full tilt. Take the way generations learned to build boats. As youngsters, they visited around the island's many boatshops — at one time 13 crowded its shores — looking, listening, analyzing; then they went home and designed and built their own toy or play fishing boats or had one made by a boatbuilder. They played at fishing with those boats until they became fishermen and built their own full-sized ones.

Isaac Beal, 57, who now manages a salmon farm, built boats for 19 years. His living room is a repository of Beals boat designs and of two- and three-masted play boats built by his grandfather and uncle. He says, "I had a fella 90-years-old tell me [when he was a boy] he waded and walked clear to Alley's Bay [on Great Wass Island], bought a play boat off'n my Uncle Floyd, and sailed it two miles back to his house." As a youngster, Isaac himself had a lot of play boats and swapped them around the way city boys swapped baseball cards.

Boats always fascinated boatbuilder Ernest

Libby Jr., 63. "I was designing boats at 10," he recalls. "I always made toy boats — play boats; towed 'em around the cove on a string." He'd go around the boatshops to see how the boats were built, and as soon as he could work a plane, hatchet and a few other tools, he began making two-foot-long sailboats and lobsterboats. Libby played with toy boats all summer long, making and setting out small trap buoys along the shore. His sons did the same thing; then the practice stopped. Last summer, though, he saw play fishing boats in the coves, so it's coming back.

Earl Faulkingham, 48, remembers having Libby build him a play boat, which he still owns. Then 10 or 11 years old, he paid for the boat by digging clams at $4 a bushel and says, "I took real good care of it, it meant so much to me: as much as a car to a teenager today." High school in Earl's day meant attending Beals High School for the last year before consolidation, for lack of students, with Jonesport High School in 1968. Years before, students from Beals had to cross Mooseabec Reach by boat to attend Jonesport High. Many parents worried about their children making the trip in rough weather, so they decided to separate from Jonesport and have their own high school. Three times a group from Beals went to Augusta to plead their case. In the end, their stubbornness paid off with the incorporation of Beals Island, in 1923, as

BRIDGET BESAW GORMAN

Boatbuilder Ernest Libby Jr. started building models at age 10.

a separate town. Islanders, not the state, built an addition to the elementary school for the high school, which graduated its first class in 1925.

Its students excelled in many ways: Norma Wilcox (now Backman) brought home the first trophy for the Washington County Public Speaking Contest in 1942; it returned again in 1952, 1955, 1957 and 1958. Beals's basketball triumphs typify the islanders' determination to succeed: in the 1950s, Beals High School won three state championships despite a home court of only 50 by 25 feet and hoops only 40 inches from the low ceiling. In

Dwight Carver

1950 the entire school had 32 students, 17 of whom were boys. Eleven of those 17 made up the team. And they won. As Junior Backman (Herman Backman Jr.) puts it, "Whatever boys there were, were on the team." In the 1970s the Jonesport-Beals team won state championships for five years straight and over the years brought home 19 gold balls.

Dwight Carver played on the championship teams of 1970–1973. He speaks of the members having "a tremendous love and respect for each other," and says that every Sunday morning he and another team member visit a third who now lives in Milbridge. "When we grew up," he explains, "everything we did, we did together from the time we were big enough to start running around the dooryard till we were 18."

By 18, most Beals boys go fishing for real. One kind of fishing predominates, lobstering, and one kind of boat, the Beals type, is designed for it, made by fishermen for fishermen.

Lobstering in the extreme waters of downeast

Maine has always been difficult. Fishermen must haul their traps in tune with the tides, which rise and fall some 30 feet — the height of a three-story building. Tides run so hard out in what they call the "main tide," away from the shore and islands, fishermen usually haul when the tides are slack: at what they call low-water slack and high-water slack. When the tide is running at full speed, it can make eight to 10 knots. Fishermen sometimes have to wait hours for the tide to slacken, especially when it's going out. At ebb tide, the force of the water drags buoys under.

Beals's engine-powered lobsterboat building started early in this century. Vernal Woodward, 91, thinks Morris Dow and Alton Rogers built torpedo-stern boats at Beals before William Frost arrived from Canada, though Frost generally gets the credit. Harold Gower and Riley, Alvin, Vinal and Floyd Beal worked with Frost, then continued on their own. George Brown worked alone. Today's boats are based on Frost's and Gower's designs.

Because Beals boats are wider than most, they have a more stable fishing platform and can hold more gear. Wider now than they used to be, Libby's 34 has a beam of 13 feet. Those made a few decades ago were about three feet narrower, and the old torpedo-stern boats weren't more than seven feet wide.

Within the general Beals Island style, designs by different builders are identifiable. "You can tell one of Calvin's by his windshield," says Osmond, 67. (It slants forward instead of back.) He adds, "Willis always puts on a longer cabin than we do," and says the bilge on his boats is rounder than the V-shape found on the other local craft. Three years ago Isaac Beal spotted an Ernest Libby Jr. boat with its high bilges, in an Alaskan boatyard and says, "Of course, I recognized it in a minute." Libby's boats, noted for their speed, are sought-after by lobsterboat racers.

The races originated on Mooseabec Reach well over a hundred years ago, back when lobsterboats carried sails, not engines. Those yearly races have done nothing to dampen the competition between Beals Islanders and anybody else. Dana Rice, of Winter Harbor, recalls, "Three days before the race there was no limit to the things they'd do to beat each other, legal or illegal." They'd plane down their boats, even break the glass on their windshields during a race to lower wind resistance. Bennie Beal, though, is the ultimate competitor: his 46-year-old oak-and-cedar STELLA ANN has roared past all younger, lighter opposition to hang onto the record for the world's fastest lobsterboat.

Today only a few boatbuilders remain on Beals: Calvin and Osmond Beal, and Ernest Libby Jr. build boats only in winter, and they're all fiberglass. Five different boat shops build Calvin's

COURTESY OF THE BANGOR DAILY NEWS (2)

Sandi Carver

designs, H & H Marine in Steuben builds Osmond's and Young Brothers in Corea builds Libby's. Willis Beal, who mainly designs and builds wooden boats, works on and off. R P Boats in Steuben builds his fiberglass designs.

Boatbuilders changed over to fiberglass because upkeep on a fiberglass hull is minimal; because a wooden hull calls for 1,000 hours of labor and one of fiberglass, 40; and because good oak and cedar were harder to come by.

In the days before there was a car ferry from the mainland to Beals, Libby says they'd have to float the boat lumber across Mooseabec Reach, the oak tied on top of the cedar so it wouldn't sink,

then they'd haul it up the bank to the boatshops. In 1958, a 1,000-foot toll bridge joined Beals to Jonesport, bringing with it accessibility; animals hitherto not seen on the island (skunks, raccoons and foxes); and increased intermarriage between Jonesporters and Beals Islanders. Some people thought the bridge would be the ruination of the island, but they were wrong. Forty years later, the summer visitor has crossed the bridge and taken up residence; but, as with the skunks and coons, Beals Islanders have learned to tolerate this new animal, too.

1999

129

Oh, Hell, Skip

Jan Adkins

All wrong. My diaphragm frosted over and an old penny appeared in my mouth. Wrong. Not just the disorientation of perspective with wind shifts, not day and night range, confusion, just wrong. Lane's Island still lay ahead, the wind still rushed over it, the boat still swung to it, but now the train of lobster cars was interposed between FOXFIRE and the island, too close, and the anchor rode disappeared into the darkness at an impossible angle, almost horizontal.

I pushed the hatch farther and peered into the inhospitable night, cold and half water, until I shivered uncontrollably, wishing childishly and devoutly that this bad thing would stop, now, and I would be back in my good bed until morning birdsounds.

Much has been made of the Tahitian navigators. Overrated by half, I say. It was warm over there, and they wore the same thing at night as they wore in the day, a yard of cloth to keep their parts out of the rigging. They didn't get up at night and put on four layers of clothes to save their bacon.

Your really intrepid explorer is a disorganized urbanite who has forgotten everything he ever learned about sailing, plagued with ghosts, desperately trying to find a working flashlight, stumbling against someone else's bulkheads without his contacts pasted over his eyeballs as he tries to find his glasses. In Shackleton's stern words, "Adventure is a sign of incompetence."

I flicked on the spreader lights. On most nights it is best not to turn the spreaders on at all. The harsh cone of light maximizes and dramatizes, an effect you do not need at 3 a.m. It opacifies the water and picks out the white beards of whitecaps from the far black; in a running sea you can feel surrounded by a nightmare. In a stormy harbor the circle of light floating against the uneasy surface emphasizes the current, the spindrift, bits of flotsam that seem insidious, caught in the night; the rain blown in salvoes through the cone forces a giddy feeling of forward rush. Crouching on the foredeck, steadied by the pulpit railing, I could follow the anchor rode from bow chock out along its shallow dip to the swivel, where

Broad Reach (detail), William Thon, 1978

JEFF DWORSKY

a length of heavy chain stretched taut against, oh shit, a Danforth Standard #135 anchor with one fluke caught in the timber corner of the right-hand lobster car.

The night, the storm, the lobster car did not dismay me, but I crouched in the bow terribly affected by the length of the rode. I had secured a 14-ton boat (on a bottom later described with a wince as "kelpy") with a length of anchor line that wouldn't have made a good dog-run. I'd anchored at dead low and run out "enough" — enough to look safe. But as the bay added 15 feet the angle became untenable, and the anchor skittered along the bottom grabbing and sliding, its jerks unfelt in the buffet of the storm.

Some perverse water spirit had, with wit and no small kindness, allowed the bow chock to nod across the very corner of the car, the spoon of the bow within inches, enough to catch the rode on the corner timber but not enough to touch. Even a city boy would have felt that. FOXFIRE backed away from the wind, the rode pullied across the corner and lifted the anchor clear of the bottom; it was (shudder) lifted out of the water, over the side, and miraculously one of its flukes sank into the spruce.

I bobbed with the dinghy, keeping station beside the miraculous anchor, looking at it again with a weak flashlight. I pulled out into the dark beyond the cars and looked back at the FOXFIRE truckstop — good spreader lights on that boat. Rowing farther out the light died quickly, gray buoys and black rocks cluttered the little cove, the water became inhospitable.

A Maine-built lobster car is good holding ground, if you can get it. Anchoring technique is a little tricky, the subject of my next article, but once you're on, you're set. I'm not saying it makes for a comfortable night: all night long I popped out of the forehatch like an overanxious Punxsutawney Phil, feeling about as much at home on the water as a mountain groundhog, beginning to believe that West Virginia boys born to it, like the Captain, out of a line of Narragansett merchant captains, born to the water, in the blood, not worth a god-damn in the mines but great on the ocean sea.

Morning disregarded my several difficulties of situation and confidence and broke sunny and brisk, as if nothing was wrong. Before the air had taken up all the color from the sun I was on deck with my crew, planning the maneuver. I could have used a set of deck apes; I had two children and a small Californian woman, but they were fierce, every one of them. The wind was still up shifting quickly now into the north. Everyone had their positions.

I was memorizing the lobster buoys around us, only one close, on my starboard quarter about five yards. We were being streamed south, almost parallel with the face of the cars, the Contender was in the dinghy, rounding the far side of the cars to dislodge the anchor, Sally and Sam had been told off to the bow where they would haul in like lightning and snub, the engine was on and in neutral, we would drift below and swing under the beneficent cars with our anchor ready to drop in the same cove (with more sensible length). She struggled with the anchor, I eased into gear and offered slack, creeping forward.

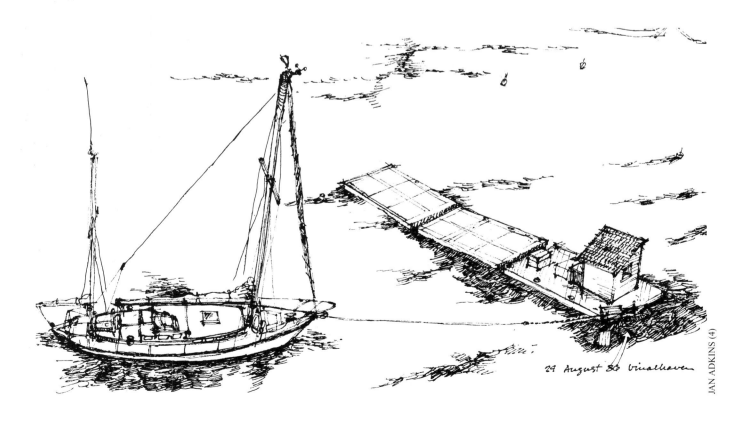

29 August 86 vinalhaven

JAN ADKINS (4)

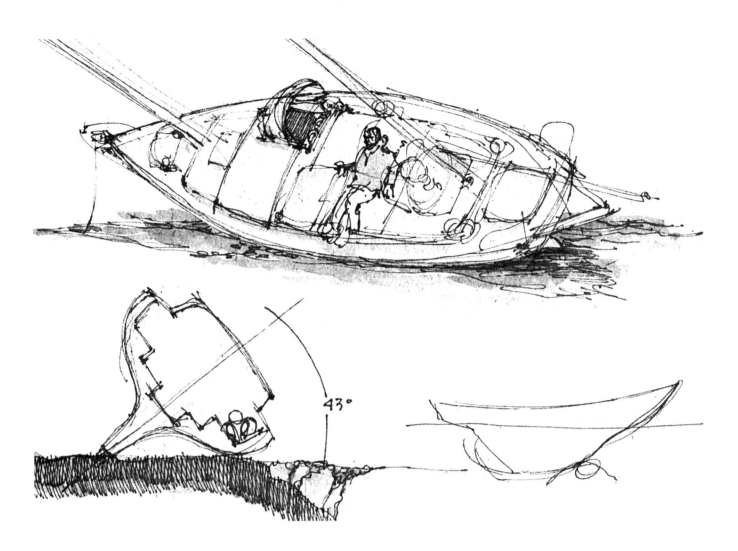

Anchor free. Hauling clear of the water. The wind is shifting in gusts, a quick look around for the inboard pot, new maneuver for new wind … pivot to starboard away from the cars and then swing under them. Starboard rudder, half ahead for a clean turn, reverse, half ahead, not answering, no response. No inboard pot. No control. Bow swings off to starboard with a new gust, off toward a ledge of rocks that looks like an old Roman wall.

Terrified and helpless as in all boat disasters, because they unfold so certainly and so slowly from a small seed of inattention, a seed germinated in the absence of that water sense that must be renewed constantly, like vows and offerings to Poseidon. Hard reverse, nothing. Sally and Sam beginning to feel a break in planning, looking uncertainly from the bow. The Contender shouting from the cars, "What are you doing?" No answer, no control.

FOXFIRE continued her swing but not her downwind progress. The wall remained a threat, 200 yards away, while we settled stern to, moored by the propeller (the subject of another technical article) to unseen lines. Unseen and untrustworthy: I had the anchor dropped at the bow and several fathoms of rode paid out against the possibility of the lines or the propeller letting go. Once again,

the water spirit had surprised then saved me with a sense of humor I was not in a mood to appreciate; and the spirit's supply of joy buzzers and wax lips was not yet exhausted.

Sunday morning. It is illegal to pull pots on Sunday in Maine. Vinalhaven fishermen, addicted to hard labor early in the morning, use this time to sort and weigh their catch from the community lobster cars moored in the harbor. The cars on which I moored for the night and near which (my transom swung ten feet off the southerly car) I was experimenting with propeller-mooring technique. Within the next 15 minutes every lobsterman in Vinalhaven was bustling about on the cars, shifting hatches and stacking baskets and avoiding looking too obviously at the yacht snagged off to leeward.

A small song of praise for the lobstermen of Vinalhaven. They work a demanding, dangerous trade; they work alone on perilous water, in conditions that compound the peril at a fearsome rate; they are watermen of consummate skill, keeping a level of knowledge and technique unattainable to anyone from Away, or anyone who does not spend life on, with, the water; they are the marks against which serious work on the water is compared. Two dozen professional watermen, a blinding blur of bright yellow bibfront. Helly Hansen's and not a guffaw among them. Not a sneer, not a slur, all opportunity for easy jibes supplanted by genuine concern.

Moreover, they went out of their way to convince me that they did the same thing all the time:

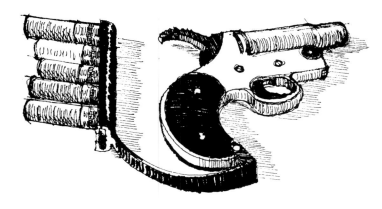

"Oh, hell, skip, you've got y'self a real mess there, you do. I've done that a time or two, I have, yessuh."

Another spoke up. "Sure have. You want to get Jimmy Knowlton out here to clean that off for you, dive down there and do it in no time. I'll get him on the radio, 'f you want."

"Well," I said, looking at the water, "I got myself into this, I figured on diving down there and clearing it up, myself."

"Oh hell, Skip," one of the lobstermen squatted down and squinted with professional skepticism under my transom. "Hell, I don't think so," he said sadly, shaking his head with professional finality.

" 'At's Wilbur's mooring you got wrapped 'round your wheel," a younger man said, not looking up from his own inquiry into the perverse nature of line under water.

Wilbur stood up and said, apologetically, "Yuh. Three-point mooring. 'At'll be a fierce tangle 'round your wheel. You let George call Jimmy. He'll fix you right up." He shook his head again, as if this treachery of line against his fellow men (me) was too much to bear, then said, "Why, you know I done this three or four times, I did. Yuh. Done it once when I was going hunting, come out here to get something and wrapped my wheel right up, never did go. Nope." New resolve, a penance. "You know what I'm gonna do? I'm gonna sink that son of a buck, yuh. 'At's what I'm gonna to do. Sink her." That would teach it, his nod said. Yuh.

Jimmy Knowlton arrived, crisply efficient, wearing Levis and an oxford shirt which disappeared as he stepped inside the folds of a black rubber dry suit, adjusted his mask and regulator, and duckwalked his fins backward over the edge of the car. Bubbles came up around FOXFIRE's waterline.

I went below to find something sharp to cut my wrists. Sam was heading for the companionway ladder with a box of Triscuits in his mitt. The thought of the freckled kid sitting on yacht cushions eating TV food while the lobstermen hoisted and the hired diver extricated, blew several circuits. I said harsh things to the boy and bid him stay below in his cabin.

In 45 minutes Jimmy had the propeller cleared and the mooring restored. We were made fast to the lobster cars, stern to, and it was time to let go. I gave the helm to the Contender and told her to bring FOXFIRE around to port as soon as I brought up the anchor and let go the last line. She did. Too fast and too reliant on the rack-and-pinion steering of her BMW. In full sight of the assembled lobstermen of Vinalhaven, and as I was leaping back into the cockpit like a roebuck leaping through brush, we ran into the nearest lobsterboat. The bobstay rode up onto his transom and we came to a nodding stop.

Lobstermen's mothers spend years teaching them to be polite: although I think I saw one waterman look away with his eyes tightly closed, there was only professional concern in 24 pairs of eyes for a moment and then they were all studiously examining objects on the other side of the harbor.

"Back down, back down!" I hissed.

But the anchor, which was hanging just below the water's surface, made a grab at the small lobster crate streaming aft of the violated boat (out of habit, probably). As we started backward the chain came taut and the crate came up out of the water and we were anchored to the lobsterboat.

"Neutral, neutral!" I hissed.

It was a moment's work with FOXFIRE's beautiful boathook to free us, and we moved off slowly, past the lobster cars. As we passed the lobstermen, Wilbur called out, "Where you gonna anchor?"

I pointed to the cove. Two or three pairs of lobstermen looked at each other as if I'd announced that I was converting to Islam. Wilbur said, hesitantly, "Kinda kelpy, ain't it?" I nodded, I'll bet it is. I took the helm and motored into the outer harbor to make careful circles, not even confident of those.

Calls on channel 16 yielded some static but no marina. It was the dapper Jimmy Knowlton, perhaps alerted by the kind lobstermen, who returned — "You need a mooring, Skip?" — and led us up the harbor to a friend's big commercial ball. I lashed down the mooring pendant with several more turns than were necessary. We were on.

Sam let go the dinghy, then.

It floated sweetly across the harbor into the shallows with its oars cocked up in the most despicably lubberly fashion, and I watched it all the way, standing, on the lazarette hatch. I could hear Sam behind me, breathing, occasionally making little sounds in his constricting throat. He was thinking, I'm too young to smoke a cigarette, so I'll only get a blindfold. The dinghy fetched up against a little dock, safe enough but mightily inconvenient. I walked forward to Sam and reached down for the centerboard winch handle.

BOY BEATEN TO DEATH WITH WINCH HANDLE BY CRAZED DAD. Page 1.

I handled the handle to Sam.

CRAZED DAD DEMANDS DUTIFUL SON TO BEAT HIM WITH WINCH HANDLE. Page 12.

I pointed to the centerboard winch. "All the way down. All the way up." I said. "Then I'll put you ashore at the dock and you can walk around and get the dinghy." One hundred and twenty turns would make him remember. I went below wishing there were a bigger, tougher winch to absolve my own guilt.

But Fraser Walker, who runs the fish plant, was on the harbor with his son in their jaunty Drascombe Lugger, and folks in little boats are always looking for something purposeful, smugglers at heart. Walker and son smuggled me ashore and the dinghy was returned. Sam was forgiven (after the turns), and someone else was forgiven.

It was a necessary removal of an obstacle but a painfully difficult effort of will: I had to forgive myself. I could not absolve myself of the responsibility or persuade myself that I had not anchored stupidly. I could, though, allow that being occasionally stupid was not inconsistent with honor or any of the workable virtues (chastity and temper-

ance elude my aspirations). Could I embrace the idea that a person who anchored stupidly might still be good at heart?

Easy to forgive someone else, much harder to offer comfort to yourself. Why? Guilt is so useful; it is the coin in which we pay debts, buy into ideas, buy out of discomfort. Dispensing with guilt is no easy matter for it has a function, propping up shaky walls in the mind, and tossing it out lightly can be dangerous. To forgive myself for anchoring on short scope in a blow I had to tiptoe around my Fake Beard guilt: gulling simple folks ashore into believing I was a sailor, pretty low considering I was raised in West Virginia. Was Joseph Conrad from West Virginia, I ask you?

They'll find out and strip my epaulettes and beard off, but until then I'm ignobly playing a shallow role with a False Beard. I also skirted the guilt of losing, or of never having, the Captain, the great bolus of angry confusion that intruded on the simplest parts of each day. Father Guilt: I am a poor example to my children; they are growing away from me, growing without guidance or wise counsel: why am I so self-conscious with them; why is Sam so dirty; why don't I feel like Bill Cosby?

So there I am, wrestling like a Laocoon with the Blue Guilts on a 41-foot yawl in Vinalhaven harbor, on a good mooring, with food and booze and a pretty lady. Grisly, right? Still, it took an hour or so to admit that the guilt was payment for something, a complex deal in emotional finance, and that giving up the guilt was giving up something else. What? Probably some agreement of ascendance with the Captain, a baring of my clumsy throat in tribute, the shreds of an old, long argument that part of me would miss, once it was gone forever. Forgiving myself for dropping the hook like a roofer would not excise it totally but I felt it going.

"I've forgiven myself." I announced to the Contender.

She smiled with relief. "We wondered how long we'd have to pussyfoot around. I forgave you hours ago."

"Ah, you see, but you don't count when it comes to forgiveness."

1989

Toy Boats

George Putz

Daily, for many years, I have walked the shores of Maine's Penobscot Bay. These walks are solace against winter, work, and the misbegotten affairs of state and heart. Almost always there are fishing boats in sight, and at the high-water line a collection of debris cast up by the tide. Now and again, among the sea junk, little toy boats made by children wash up on the beach. Some of them are encrusted with barnacles and seaweed, their paint long gone except where it was trapped under the heads of nails and screws. Each year produces at least a couple of them.

In 15 years of regularly walking the shore, my collection tallies at around 30 of these witnesses to emerging skill, good counsel, hard work, enjoyment and loss. It's the good counsel and loss combined that strike me most. My neighborhood is an especially nautical and maritime one; the youngsters around here are knowledgeable about boats and handy with tools. Some of these lost creations are perfectly astounding in their fidelity to both reality and genius — unique to their creator, but unmistakably fishing boats. In an age of toy-store plastic, they are powerful and primeval, brought forth from Daddy's woodscrap pile, nail pouch, and by now misplaced hammer.

Remember those times at ditch and shore? How we were actually in those boats, doing boat things; carrying cargo, towing other boats, going really fast, winning naval victories? We built the craft, painted it, used it, and then, and then …. Something happened to it. The string broke. The

PETER RALSTON

ISLESBORO HISTORICAL SOCIETY

My neighborhood is an especially nautical and maritime one; the youngsters around here are knowledgeable about boats and handy with tools.

dog ran off with it. The tide rose. We forgot.

These shores I walk are remote, almost never visited by children (or anyone else). They are surrounded by water with powerful flows of tide and current. The chances of a lost toy boat landing on them are not good — and yet there are so many — so many inadvertent Paddle-to-the-Seas. These chunks of wood, with their crude working forms, are reason for optimism: their creators were intrepid enough to lose their creations. Unlaunched ships don't come in; to give our small works to the vast works of the world is a proper thing.

The lovely boats are not simply artifacts of childhood or lessons in pathos. They remind us that somewhere, for our own wither-away creations, a beach, a landing place, a walker-by-the-water is waiting.

1985

137

Teels Island, Andrew Wyeth, 1954

The Soul of a Work Boat

Mike Brown

The soul of my first boat appeared 20 years before I was born. Silently, a cedar seed dropped from a conifer forest on the south side of Bird Hill in Northport, Maine. Fate ushered the myriad predators of cedar seed past my boat-to-be until the warm rain of autumn gave it legs to penetrate the needle carpet of the forest floor. Come spring, my boat-tree was born.

It would be four and twenty years before Oscar Drinkwater would fell my boat with his crosscut saw and double-bit ax and haul it with a team of horses named Ike and Bite to the hillside road banking. There, my tree-boat joined others on their journey to Dan Robinson's mill to be sawn into layered boat boards. My father, the fisherman, had said one night sitting on our herring weir pocket frame, one night when the fish shimmered in rare abundance in the heart of our weir, my father said that night that every boy fisherman needed a boat of his own. I was six years old. The Bird Hill cedar eventually came to Harold Martin's boatshop on the Belfast waterfront. And there, in the corner by the double drum stove, my father would work on my boat when the herring were not running. She was a punt, she was. Higher

than my Gram's apple pie and looking about the same, pointy stem to meat loaf stern.

My memories of my punt a'building are smothered in the sight and smell of cedar shavings. On Saturday afternoons, tide and fish permitting, my dad and I would walk to the boatshop along the shore from our house. It was not the quickest and easiest way of getting to town for fishermen who didn't own a car. But it was the familiar character of the shore that made the five miles more tolerable. We would tote a supper of biscuit sandwiches, shortbread, and jam dessert. Drink was strong coffee from a blue agate pot atop the oil drum stove. The late afternoon walk and biscuits made for a sleepy boy. I would curl up back of the stove in a boy nest of cedar shavings and wake only when the calloused hand of father fisherman shook me. My punt was completed

between such naps.

We rowed her home, my punt, my dad, and me. On a going river tide. Punt foreman Harold gave me a christening gift. A cedar scoop. On her maiden voyage along the Battery Shore, I thought my punt was doomed. I scooped until I had blisters. But she came to life, my punt did, and swelled her seams shut with all the subtle grace and power she used growing tall and straight on the south side of Bird Hill.

I had a punt with no name for years. She just finally wore out. One fall day my dad said punt wouldn't make it through another fishing year. We burned her that night on the Northport shore. In the morning I gathered her crematory bones of clenched nails and dug a hole by the big rock and buried them where they are today.

1990

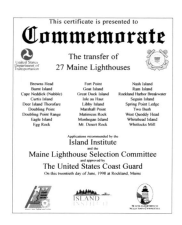

This certificate is presented to

Commemorate

The transfer of
27 Maine Lighthouses

Browns Head	Fort Point	Nash Island
Burnt Island	Goat Island	Ram Island
Cape Neddick (Nubble)	Great Duck Island	Rockland Harbor Breakwater
Curtis Island	Isle au Haut	Seguin Island
Deer Island Thorofare	Libby Island	Spring Point Ledge
Doubling Point	Marshall Point	Two Bush
Doubling Point Range	Matinicus Rock	West Quoddy Head
Eagle Island	Monhegan Island	Whitehead Island
Egg Rock	Mt. Desert Rock	Whitlocks Mill

Applications recommended by the
Island Institute
and the
Maine Lighthouse Selection Committee
and approved by
The United States Coast Guard
On this twentieth day of June, 1998 at Rockland, Maine

Giving Away
the Lights

Steve Cartwright

Astride the busy channel into Boothbay Harbor lies a quiet, natural oasis: five-acre Burnt Island, home to wild berries, birds and a lighthouse with a 179-year-old stone tower, the second-oldest original lighthouse in Maine. One of the last manned light stations in the country, Burnt Island was fully automated in 1988, when the last family to live there pushed off for the mainland. The old house grew musty, the roof sprang a leak. Weeds took over the yard where once the keeper's children played.

Until Elaine Jones came along. "I'm the kind of person who says 'I'll find a way,'" said Jones, 44, who — on behalf of her employer, the Maine Department of Marine Resources (DMR) — won custody of Burnt Island Light over a competing proposal from a group that had leased the property since July 1992. Jones is confidently planning a living history museum and unrestricted public access to the five-acre island.

She doesn't just talk, she rolls up her sleeves and scrapes and paints, and she has a knack for persuading others to pitch in, whether the matter at hand is cutting brush or donating funds and fur-

nishings for the keeper's house.

Through the Island Institute's Maine Lights Program, the deed to Burnt Island was transferred from the U.S. Coast Guard to the DMR, which maintains a facility at McKown Point, just a brief boat ride from the lighthouse and island it now owns. Jones, a tenacious advocate for education, preservation and public access, is using volunteers to turn Burnt Island Light Station

PETER RALSTON

into a light keepers museum, complete with period-dressed staff and overnight accommodations for Maine students. "I'd live here year-round if I didn't have a young family," she said.

Jones has located former light keepers and their descendants, such as the octogenarian great-granddaughters of keeper James McCobb (1868–1880), who have agreed to help with restoration details. "I'm an educator," Jones said. "To see the kids [who visit the light] react is moti-

vation. Burnt Island is a legacy I'm leaving to the people of the State of Maine and the country."

Previously, Jones built another educational facility for DMR, a public saltwater aquarium at McKown Point where children can see and touch the fish.

Burnt Island is just one example of how 36 coastal stations in the congressionally-approved Maine Lights Program are being restored and returned to the people of Maine. These lighthouses must be maintained to high standards and the grounds kept open to the public.

The Maine Lights Program officially ended in the fall of 1998, when title to 27 of the 36 lights were awarded to new owners. The remaining stations failed to attract support because of remote location, lack of land or hazardous access.

Anne Webster-Wallace, who for two years directed the daily activities of the Maine Lights Program for the Island Institute, said she hopes proposed legislation now being drafted by lighthouse preservationists will assure that these remaining Coast Guard lighthouses stay open to the public.

In all, Maine has 64 coastal lighthouses plus a small stone lighthouse on Cobbosseecontee Lake near Augusta. Some of these are already privately owned. Webster-Wallace, who lives on Georgetown Island, brought to Maine Lights her experience organizing Friends of Seguin Island Light, a station that dates to 1795 and was recently deeded to that group. "One thing I didn't realize would be helpful," she said, "was that I had 12 years in real estate. It really did save us a lot of legal fees." Nevertheless, the Island Institute was obliged to invest $300,000 in the effort, largely

Iris at Sea, Jamie Wyeth, 1994

legal fees and title research.

While some "non-program" lighthouses remain in limbo, others have new owners but await needed maintenance and restoration. "Some things were more agonizing than they needed to be, but in general it was very successful," Webster-Wallace

said. "The applicants took the responsibility very seriously."

Burnt Island is one of the most exciting success stories among many that grew out of the Maine Lights Program. As in all cases, the deed changed hands through the work of a five-member selection committee whose members included Vinalhaven Town Manager Susan Lessard, who lives at Brown's Head Light, which, thanks to the program, is now town property.

For seven years she has lived at the 1832 light station as a perk of her job. "It's a wonderful, 150-year-old house," she said. "It's a 270-degree water view ... eight months of the year it's an isolated location. You get quite proprietary about it. I don't have to light the lamp [which was automated in 1987] but I do mow the grass and pick up the trash."

Lessard, who heads a new coast-wide lighthouse owners' group called Maine Lights Inc., speaks highly of islanders. "They judge you entirely on your work ethic," she says. "This community is a good fit."

She must share her homestead with summertime picnickers and picture-taking tourists, and even puts up with romantic proposals when published stories mention she is single. She said those who propose to her don't know she sometimes lugs water through biting winter wind when her "interesting" water system fails. "The word 'romantic' isn't the first word I'd use to describe the place. It's a physically demanding life."

There were no takers for Franklin Island Light in Muscongus Bay, a bare tower on an island already protected as a bird sanctuary. The same was true for Halfway Rock, a barren ledge between Seguin Island Light and Portland Head Light.

Ultimately there was no formal interest in Manana Island fog signal with its soaring repair estimates; Wood Island Light, near Old Orchard Beach, is still not spoken for; neither is Goose Rocks Light, a "caisson" light on a ledge between Vinalhaven and North Haven. No one took custody of Ram Island Ledge, a lighthouse in Casco Bay that literally has no land base when the tide's high.

Two downeast lights have found no friends, at least not yet. Little River Light, at the entrance to Cutler Harbor, has plenty of potential but had no serious takers, according to Webster-Wallace. The same goes for Moose Peak, an 1827 lighthouse on Mistake Island near Jonesport.

Located at the southwest corner of Penobscot Bay is Whitehead Island Light, which takes its name from a highly visible granite outcropping. First commissioned by Thomas Jefferson, it marks the Mussel Ridge Channel, a major highway in the age of sail and still a busy passage. The original

PETER RALSTON

Curtis Island

1803 stone tower at Whitehead was rebuilt in 1852, and one of two keepers' houses still stands. The other was demolished by the Coast Guard many years ago.

Through the program, Whitehead Light is now the property of Pine Island Camp, a boys' camp founded 97 years ago on Pine Island in the Belgrade Lakes. Pine Island now owns the keeper's house with its panoramic view — on clear days, anyway, since Whitehead is notoriously foggy. The camp also owns the granite light tower itself, while the Coast Guard retains rights to operate and maintain the actual light and a nearby foghorn as aids to navigation. This is the typical arrangement for transfer of active lighthouses. The Coast Guard needs no access to discontinued lighthouses, although sometimes these are illuminated with plain bulbs for aesthetic reasons.

Third-generation Pine Island Camp director Ben Swan wasted no time in applying for the light, which shares Whitehead Island with a former life-saving station built in the 1880s — before there was a Coast Guard. The camp and the Swan family have owned most of Whitehead Island for the past 30 years, renovating the lifesaving station and a Coast Guard barracks in the 1970s.

Renovations to the keeper's house are already under way, much of it done by Pine Island youth, and more work is scheduled for summer 2000.

Swan, 44, praised the Maine Lights Program's Selection Committee, especially the role played by its chairman, retired Coast Guard Rear Admiral Richard Rybacki of Falmouth, and the steadfast work of Maine Lights director Webster-Wallace.

"It was, for once, things the way they ought to be — almost more than any other applicant, we knew what we were getting into. We were there, and we knew what to do. Every time we needed someone, there was someone there," Swan says. Help has come from people like carpenter Nick Buck, a former Coast Guardsman who was stationed at Whitehead in 1976 and 1977. He now heads the camp's Whitehead Lightkeepers program: two three-week sessions offering an island work experience for teenagers. David Gamage, who owns a cottage on Whitehead built by his lightkeeper grandfather, has also pitched in, researching renovation work.

Whitehead is open to visitors, and Swan is grateful the island isn't inundated by boaters. "It's an island and it's not that easy to get to, and we're thankful for that," Swan said. "We feel very fortunate that we are able to manage it." He said Pine Island's four-day visits to Whitehead can be a highlight in a camper's summer.

Webster-Wallace believes lighthouses can be a special experience for people of all ages. She is particularly pleased that the Maine Lights Program

COURTESY OF ELAINE JONES

Elaine Jones

was able to present deeds to groups that already had proven themselves good stewards through existing leases with the Coast Guard. That honor roll of lighthouses includes Cape Neddick, built at York Beach in 1879; the 1833 Goat Island Light at Cape Porpoise; Seguin Island, which dates to the late 1700s; Ram Island at Boothbay Harbor; Marshall Point museum at Port Clyde; Monhegan Island museum; Curtis Island at the entrance to Camden Harbor; Brown's Head on Vinalhaven; Fort Point at the mouth of the Penobscot River; and West Quoddy Head, furthest Maine light downeast.

Admiral Rybacki, who chaired the selection panel, shared Webster-Wallace's enthusiasm. "I'm very pleased with what I hear and see of the results of the work of the committee. The legislation that authorized this process was one of the finest pieces of legislation we've gotten to work with, particularly for the disposal of property." He praised the involvement of Mainers in that process. "It was not a group of people from Washington, D.C."

Rybacki said he wasn't surprised several of the lights failed to attract support. Ease of access, the presence of a keeper's house and the amount of land were important considerations for Maine Lights applicants. But one of the rejects is a personal favorite: Wood Island Light, not far from his home. "It's one I would certainly like to own if I had had the wherewithal," he said.

Graceful and purposeful, lighthouses are literal aids to navigation, and figurative symbols of guidance; of finding one's bearings and setting a course for safe harbor. For all the affection lavished on them, many Maine stations have been victims of neglect and newer technologies.

It took the near loss — in a 1989 fire caused by an electrical short circuit — of unmanned Heron Neck Light on Green's Island in Penobscot Bay, to galvanize action to save remaining lights. At first, rescue efforts focused only on the burnt-out keeper's house, since restored. But Heron Neck sounded the alarm that led to the drafting of the Maine Lights program. Without Island Institute intervention, Heron Neck "would be a hole in the ground," said Peter Ralston, executive vice president at the Institute.

Ralston and Ted Dernago, chief of rural property for the Coast Guard, worked closely on the Heron Neck rescue. Then, half in jest, Ralston asked Dernago if it would be possible to save some more lighthouses.

"How many more do you want?" Dernago asked. And as they discussed the idea, their lighthearted conversation turned to serious planning. "The synergism between Peter Ralston and myself just exploded," Dernago said. "That was the building of a tremendous friendship that continues today."

Dernago is sensitive to criticism of the Coast Guard's role in the upkeep of lighthouses over the years. "From the Coast Guard side, we'd been maintaining and protecting these lights for years. The Coast Guard has always tried to do its very best with what little it had."

Some have criticized government officials for failing to value lighthouses. Ralston said that the agency isn't the bad guy. "Technology has blown right past these lights. To a certain pragmatic extent, they are redundant, with new navigational aids such as radar, GPS, loran."

To preservationists, lighthouses are vital beacons woven into the culture and history of the coast.

2000

PETER RALSTON

WATER

It was a Maine lobster town —
Each morning boatloads of hands
Pushed off for granite
Quarries on islands,

And left dozens of bleak
White frame houses stuck
Like oyster shells
On a hill or rock,

And below us, the sea lapped
The raw little match-stick
Mazes of a weir,
Where the fish for bait were trapped.

Remember? We sat on a slab of rock.
From this distance in time, it seems the
color
Of iris, rotting and turning purpler,

But it was only the usual gray rock
Turning the usual green
When drenched by the sea.

The sea drenched the rock
At our feet all day,
And kept tearing away
Flake after flake.

One night you dreamed
You were a mermaid clinging to a wharf-
pile,
And trying to pull
Off the barnacles with your hands.

We wished our two souls
Might return like gulls
To the rock. In the end,
The water was too cold for us.

Robert Lowell, *1993*

145

Working Waterfront *Turns 10*

David D. Platt

L ate in 1992 in the Island Institute's old offices on Ocean Street in Rockland, an informal group began planning a new publication. The Institute's first newspaper, *Island News*, had already become *Inter-Island News*, a post office staple in the 15 year-round island communities and a forum for lively discussions of ferry service, solid waste and community life, along with a generous dose of "us vs. them" opinions.

The meetings that took place that winter were focused on something entirely different: a newspaper of broader interest, greater reach and more mainland potential than *Inter-Island News*; a journalistic enterprise that would be of interest, we hoped, to anyone who depended on the "working" waterfront to earn his or her living. The health of such places along the Maine coast — and their dwindling extent,

PHOTOGRAPH COURTESY OF JAMES ACHESON

due to development and gentrification — had been documented in early editions of *Island Journal* and had even been the object of a study by the State Planning Office. (The study's most startling finding: the number of miles of working waterfront on Maine's coast in the early 1990s was less than 25, and dropping.)

Working Waterfront was actually part of a larger effort to set up a marine resources program at the Institute. Philip Conkling (who chaired the aforementioned planning meetings) notes that this initiative was made possible through the generosity of the Birch Cove Foundation, which made resources available to the Institute after an unsuccessful effort to contribute to an island-acquisition effort by The Nature Conservancy. "I'm sorry, sir,"

someone there had told Birch Cove's principal donor when he offered to "buy" a few puffin burrows in support of The Conservancy's apparently too-successful island-saving fund raiser, "but the burrows are all sold." The donor switched allegiance to the Institute and its marine resources project, and the rest is history.

The early 1990s were a good time for new publications. Desktop computers, moderate paper prices and competition among job printers had made it possible for a single individual — or, at least, a very small number of people — to write, edit, lay out, print and distribute a newspaper at reasonable cost. If you could sell enough advertising, you could recoup some or all of your costs. In the case of nonprofit groups, a little-known fact even helped keep the postal rates down: in 1993, at least, the largest nonprofit bulk mailer in the country was the Catholic Church. Will Congress raise the rates or won't it? Go figure….

Charlie Oldham

The name of the publication, it was decided, would be *Working Waterfront*. Like *Island Journal* (but unlike *Inter-Island News*) it would make use of coastal Maine's ample pool of free-lance writers, photographers and graphic artists to fill its pages. For starters, at least, the editor would design the paper himself on a computer, using the PageMaker program. It would appear six times a year, alternating with *Inter-Island News*. Thinking big, we decided to mail it to every box holder on a list of coastal and island towns, selected because they had working harbors or because a significant number of licensed working fishermen lived there. All Institute members would get the paper, as would others on a mailing list that included decision makers and others with an interest in the coast. The press run for the first issue added up to over 20,000 copies — bigger than any other "alternative" paper in the state, very big for a little place like the Island Institute; very new and adventurous for a publisher accustomed to high-end work like the *Island Journal* or a little community paper like

Inter-Island News.

What to cover? Jump-starting a newspaper is harder than it looks. We went in search of stories that might lead us to other stories: the risks of oil spills, growing nori, a profile of a Portland wharf. We sought out op-ed pieces, looked around for interesting people to profile. We made a practice, in those early issues, of building each paper on a theme: old buildings with possibilities, small-scale aquaculture, the land–sea connection, finance, real estate, the promises of politicians. We also went looking for advertisers: marine contractors, gear suppliers, small manufacturers, folks with marine-related property to sell. Building an advertising base that would support a reasonable portion of our costs (this is a nonprofit enterprise) took about a year. Since 1994 we have tried, mostly successfully, to keep the "paid" portion of the paper at about 30 percent of the total.

Ed Myers made his first appearance in September 1993, with a column called "All at Sea." He dissed a few lawyers, waxed eloquent about an aspect of aquaculture and entertained us with statistics. It marked the beginning of a relationship that lasted until his death nine years and 40 columns later. Roger Duncan followed, as did Rusty Warren and Phil Crossman. Ted Spurling had been writing for *Inter-Island News* and continued there until we merged the two newspapers in July 1997. He now writes a piece every other month, as do the other columnists.

Mike Herbert

PETER RALSTON (3)

David Platt

Good writing has been at the heart of *Working Waterfront*'s success. From the start we drew on the talents of experienced freelance writers: Christine Kukka, Francie King, Bob Moore, Bob Gustafson, Muriel Hendrix, Sandra Dinsmore, Bonnie Mowery-Oldham and others. Institute staff members, particularly Philip Conkling and Ben Neal, have made major contributions, and a new generation of Island Institute Fellows now produces a steady stream of stories. After the first few issues, responsibility for the paper's design and production passed from the editor to a professional, first Anne Parrent, then the indefatigable Charlie Oldham. Anne (and before her, Sharon Smalley) sold ads between issues; then came Sandra Smith, followed by today's real pro, Mike Herbert.

The paper's mix of island and coastal news, business stories, profiles, history, commentary, letters, reviews, marine science and fishermen's wisdom seemed to work. Circulation reached 50,000 two years ago. (We cut it back to just over 40,000 last year as the recession pushed ad revenues down, but we're still one of Maine's largest newspapers.) We inaugurated a web edition of the paper, www.workingwaterfront.com.

The public response to the paper has been overwhelmingly positive since the beginning. Last winter a survey of the Institute's programs gave *Working Waterfront* high ratings, along with the Institute's Fellows program. The lessons of all this: start small, stay relatively modest, publish as frequently as you can afford, pay attention to your supporters and advertisers, listen to your writers and your readers, and stay focused on the topics you have set out to cover. In our case the topics — the communities, economies and people of the coast of Maine — are of more than passing interest to a wide audience. Ten years into this project, we're feeling pretty good about it.

2003

CHAPTER SIX

Art of the Islands

IN THE SUMMER OF 2000, the Farnsworth Art Museum mounted an ambitious exhibit entitled "On Island: A Century of Continuity and Change." The show highlighted the juxtapositions so evident on islands: land and sea, man and nature, wild and tamed. Artists have always found such contrasts stimulating and enriching, and Maine islands have produced more than their share of great artistic expression.

"All good art is a search for meaning," writes Maine art critic Edgar Allen Beem. "What artists find on Maine's islands depends not only on what they are looking for — privacy, inspiration, natural beauty, society — but also on the sensibility of the artist and the character of the island...."

Island Journal has celebrated the visual arts from its earliest years, regularly publishing folios of paintings that reflect the strengths and special qualities of islands. In essays, art critics, poets and others have explored the meaning of the art that for a century and a half has emanated from island settings. Art — particularly painting — is only one medium for expressing the meaning of island life and landscapes, but it can be very powerful.

Photography, of course, is another. The work of Peter Ralston has been closely identified with *Island Journal*, most notably on its covers, from the start. So has that of Jeff Dworsky. "Only an island community grown to tell the hard and redeeming truths of island life," writes poet Philip Booth of Dworksy and Isle au Haut; "only an islander grown all-but-native could know how to open his camera in non-invasive ways to the faces and lives of the people he lives and works with, and cares about deeply."

David D. Platt

Beach Flowers # 2, Fairfield Porter, 1972
Above*: Spring Islands* (detail), Eric Hopkins, 1988

Airborne, Andrew Wyeth, 1996

The Art of Island Maine

Edgar Allen Beem

In a state rich with art history, Maine's islands have proven to be both haven and holy land to generations of artists. In the beginning, they came for the scenery — raw landscapes still being sculpted by the hands of God and Nature; but increasingly in the supersonic age Maine islands became a source of sanctuary — places where human life achieved dramatic focus as it was lived in isolation and open to the elements.

Since World War II, as wave after wave of "ism" has swept over American art, Maine islands have provided safe harbors to independent artists fleeing the orthodoxy of the day. And they have made both their art and these islands their own.

"Islands, by definition," wrote Alan Gussow in his landmark *A Sense of Place — The Artist and the American Land*, "are places unto themselves, but Monhegan is also a place within me — finite, singular, entire."

Alan Gussow first came to Monhegan in 1949…. "For all the distances one sees on Monhegan," Gussow says, "I have this feeling of locating things at my feet." The sea, the sky, and the island light are constantly changing, he notes, "but the rocks have really been immutable."

A strong naturalistic strain has run through the rock-hard realism of Monhegan art, from Rockwell Kent on down through marine and landscape painters such as Jay Hall

Above: *The Herring Net,* Winslow Homer, 1885

Maine Coast, Rockwell Kent, 1907

Connaway, Andrew Winter, James Fitzgerald and Don Stone. But having been colonized primarily by New York artists, Monhegan was also one of the few places along the coast of Maine where serious abstract and expressionist art took hold.

Abstraction is an urban impulse, an internalizing of experience beyond perception. In the paintings of the late Michael Loew, for instance, Monhegan is all but unrecognizable, the light and forms of the island having been transformed into a geometry of pure color. Hyde Solomon, Murray Hantman, and William McCartin are among the other accomplished painters who have rendered the island abstract. But if one knows how to look or what to look for, a place as powerfully real as Monhegan is sure to leave its traces. Such is the case with the Monhegan art of William Manning.

Bill Manning, a Portland artist who paints on Monhegan each summer, creates elaborate, constructed paintings which are essentially idealizations of natural phenomena. His abstract lines and shapes and shadings, beautiful even without appeal to nature, are like the residue of reality, the evanescent line and light of cloud shadows, ocean currents and the contrails of jets heading out over the

North Atlantic. One would think Manning could find some material anywhere, yet Monhegan remains the foundation of his art.

"It's like church to me," says Manning of Monhegan. "It's a very mystical feeling. I just have to go there."

The expressionist urge is driven by emotional needs, characterized by the free use of color and best made manifest in the human figure. The actor and artist Zero Mostel, for instance, was a colorful Monhegan summer resident who often starred in his own Picasso-esque compositions. And John Hultberg, one of the New York art stars of the 1950s, brought his own surreal vision with him when he arrived on Monhegan in 1962. Hultberg is a painter of apocalypse and decay and he loads his brush with torment.

Other painters on Monhegan of an expressionist bent trade in high-keyed color without Hultberg's sense of existential angst. Lynn Drexler, Hans Moller, and Elena Jahn all treat the island to colors it has never worn, but they do so out of excitement rather than anguish. But perhaps the most soulful and individualistic painter to work on Monhegan in modern times was Joseph DeMartini.

Described by his friend Bill McCartin as "a man of great simplicity and honesty," DeMartini painted a lonely, human universe from his rented studio on Fish Beach and he did so, most often, in somber tones of black and white and gray.

One island, many realities. Currently, the most popular version of Monhegan is painted by Jamie Wyeth, son of Andrew Wyeth and grandson of N. C. Wyeth. Living and working in a house built by Rockwell Kent, Wyeth paints a gothic Monhegan, both beautiful and strange. He has inherited the Wyeth narrative gift, the ability to transform the everyday into the extraordinary. Wyeth's portraits of island boy Orca Bates, for example, capture the native wildness of both the island and the islander.

"Orca is a true island child — at home with gulls, rocks and seaweed," Wyeth has said. "He was born on Manana, the treeless rock behind him. He belongs to it; he is part of that place."

Becoming part of Monhegan has become increasingly difficult for artists as the island's popularity (and therefore its real estate values and rental fees) increases. To give more artists exposure to its rare, deep sea beauties, fiber artist Robert Semple turned his Monhegan cottage, Carina House, over to the Farnsworth Museum in Rockland to administer as an artists' retreat. Marguerite Robichaux, a painter from the

Carrabassett Valley in western Maine who spent six weeks at Carina House in 1990, says she has returned on her own each summer since. Such is the magnetism of Monhegan.

"I live in the woods of Maine for my solitude," says Robichaux, "but I am drawn to the edges of the world, and Monhegan is definitely an edge of the world."

"One of the great moments being on an island," says painter David Little, "is that brief interval when you're on the edge of the woods, the light coming through, then you step out onto the shore and it's in your face."

David Little is a second-generation artist on Great Cranberry Island just off Mount Desert. Great Cranberry is a charmed, often overlooked island, not so dramatic as Monhegan, but in its tranquility and repose just as attractive to artists. Little inherited his cottage on the island from his uncle William Kienbusch, perhaps the most forceful abstract painter to work the peaceable shores of Maine.

In many ways the Cranberry Isles are charmed places, intimate and overgrown now that sheep and farming no longer keep the land open, but facing the mountains of Mount Desert, they own some of the best views in Maine. Artists are not so numerous on Cranberry as deer, but those who come are

Orca in Winter, Jamie Wyeth, 1990

Perry Creek and the Thorofare, Eric Hopkins, 1995

often bound in tight little coteries of friendship and family.

The first artist on the island was printmaker Charles Wadsworth, a conscientious objector during World War II who came to the Cranberries in 1946 to escape a world in turmoil. When he reported the serenity of Great Cranberry to friends back in New York, several artists who had studied with Wadsworth at the innovative American Peoples School retreated to the island as well, among them Gretna Campbell, Carl Nelson, and John Heliker.

Heliker, now the grand old man of Cranberry, is a master of the gentle impressionist landscape, painting intimate views of island gardens and interiors that speak of civility and the cultured life. He shares his Cranberry farmhouse with painter Robert LaHotan, an artist more apt to paint studio still-lifes than island landscapes, but LaHotan credits his time on the island with being a creative catalyst.

"Being here," LaHotan has said of Cranberry, "releases a lot of memories of other places even if I'm not painting the island."

In like manner, William Kienbusch used to refer to the trek from New York to the Cranberries as "climbing Mt. Everest." It was a trip downeast he made in stages, stopping off in places like Portland to decompress from urban pressures so that by the time he reached far Cranberry he was

so psyched up he'd set to work immediately recording vivid impressions of subtle island phenomena such as the sound of a gong buoy or a glimpse of the sea in vigorous, abstract strokes.

Where Kienbusch reacted to Cranberry with an excited eye, his nephew David Little conveys a deep sense of the island's stillness and calm in his delicate landscapes. For all the peace and freedom Little finds on Cranberry, however, he also believes that "an element of fear" enters quietly, darkly into island art, island life. This fear is a function of the heightened awareness of isolation and vulnerability one feels on an island, but it is also part of the beauty of the place.

Another of the Cranberry Isle connections is Cooper Union art school in New York City, where island artists Gretna Campbell, her son Henry Finkelstein, Finkelstein's wife Carleen LeVander, Ashley Bryan, Emily Nelligan, and her husband, Marvin Bileck, all studied. Though this common aesthetic bond imparted no particular artistic point of view, Henry Finkelstein had followed in his late mother's footsteps, painting closed Cranberry landscapes in vibrant colors and flashing strokes.

Emily Nelligan, on the other hand, has been an almost ghostly presence on the island, haunting the tidal marsh at Fish Point in fog and at first and last light to create exquisite charcoal drawings that reduce Cranberry to its absolute simplest terms. It is remarkable that someone who spends so little time on the island is capable of experiencing it so deeply.

In truth and in fact, of course, most of the artists who work on Maine islands are summer visitors. Of the Cranberry artists, only illustrator Ashley Bryan is a year-round resident. More rare still on Maine islands are native artists. Indeed, Eric Hopkins of North Haven may well be the only native Maine island artist of real distinction.

North Haven is a summer enclave of old Boston families, and the vast majority of the island land is now in out-of-state hands. Eric Hopkins, however, owns the landscape. His sweeping, purified aerial and marine views of North Haven and the neighboring islands of Penobscot Bay possess a kind of cosmic charm, their tilting vistas and arcing horizons politely reminding viewers that Vacationland Maine, for all its early beauty, is but a fleck in a vast, mysterious universe.

Just fathoming the mysteries of a Maine summer colony proves too much for most artists, it seems. Neither North Haven nor Islesboro have incubated much in the way of modern art. Islesboro, with its aristocratic Dark Harbor community, is Maine's society island, and most of the artists who work there are related to one another.

Decade Autoportrait, Robert Indiana, 1982

Brita Holmquist has absorbed the natural rhythms of the island and the bay into her sunny, felicitous paintings, but for many years she painted nothing but the garden view from the porch of her family's Dark Harbor summer estate. Brita is the daughter of painter Ibbie Holmquist, and from time to time several members of the Holmquist clan have exhibited together at the Islesboro Historical Society. The present younger generation includes Ian Kats, Bayard Hollins, and Joshua Outerbridge.

The most accomplished outsiders who have painted Islesboro in recent years are Hearne Pardee and Gina Werfel. Husband and wife, Pardee and Werfel are latter-day Fauves who paint dashing landscapes that are as much about the act of painting as about the act of seeing or anything seen. In this, they are aesthetic kin to Henry Finkelstein, who is represented by the same New York Gallery as Werfel. In the small, tight world that is Maine island art, therefore, it is not surprising to find that Werfel and Pardee have also spent time painting on fair Cranberry.

To find real mavericks on the Maine art scene you have to travel to a working island: the harsh, unlovely, quarry-pocketed, magnificent island of Vinalhaven to be exact. Vinalhaven is a hardscrabble, make-do place, and the artists who work there are something like hermit crabs, seeking shelter in the abandoned. The most noted of these island exiles is Robert Indiana.

Robert Indiana, one of the stars of Pop Art in the 1960s, first came to Vinalhaven in 1969 at the invitation of photographer Eliot Elisophon and since 1978 has made his permanent home in the former Star of Hope Oddfellows Hall. Famous for his LOVE paintings and sculptures, Indiana is an artist who deals in signs and symbols, so Vinalhaven manifests itself in his work in emblem-

Cold Ruffles II, Brita Holmquist, 1989

atic rather than visual ways.

His greatest achievement since moving to the island has been Indiana's "Hartley Elegies," a series of paintings and silkscreen prints made in homage to Marsden Hartley. Hartley, a Lewiston native who became one of the giants of modernism, spent the summer of 1938 on Vinalhaven, but Indiana's "Hartley Elegies" are based on the symbolic paintings Hartley did during World War I to memorialize a fallen German officer friend.

During the 1980s, Vinalhaven became something of a printmaking mecca after Pat Nick rented the former island schoolhouse and installed her Vinalhaven Press there. Indiana was one of the first artists to make prints at the press, but the Vinalhaven Press's chief modus operandi has been to entice established and emerging artists up from New York with the prospect of a summer escape to a Maine island. While on the island, the artists execute prints in collaboration with master printmakers, and then Nick markets them during the winter back in New York.

Though the Vinalhaven Press has no overriding look or style, the artists Nick has handpicked have tended to be tough-minded individuals with gritty aesthetics that fit the nature of the island. Among the artists who have worked at the press are islanders Indiana, Peter Bodnar, and Carolyn Brady, and outsiders Louisa Chase, Susan Crile, Robert Cummin, Leon Golub, Charles Hewitt, Robert Morris, and the Russian émigré team of Komar and Melamid. Charlie Hewitt, an Auburn, Maine, native, has also teamed up with two other outstanding Maine artists who summer on Vinalhaven, Alison Hildreth and Katerina Weslien, to make prints independently.

Another refugee from the mainland art wars is Tom Lieber, a California painter who first came to Vinalhaven to visit artists George Bartko and Peter Bodnar. Lieber fell in love with the island and now paints summers in what once was the Vinalhaven Knights of Pythias ballroom.

Lieber's abstract paintings merge elements of the human figure and elements of nature into luminous internal landscapes that have an aura of spirituality about them.

"Being in the middle of the Atlantic Ocean is a real cleansing experience for me," says Lieber. "Just being on the water, seeing the reflections come into my head through my eyes, is an education for me. The fact that I'm on an island surrounded by water, even though there's solid ground beneath my feet, has a certain quality of vastness that enters into my work, making it more special."

The island experience is one of both physical and psychic removal, as painter Larry Hayden learned in 1979 when he moved out to Peaks Island. Peaks is a blue-collar island just a short ferry commute from downtown Portland, and during the 1970s and 1980s several artists — among them Ellen Gutenkunst, Hayden, Biff Higgison, Richard Hutchins, Chake Kavookjian, Patrick Plourde, and Claudia Whitman — moved there, attracted by the beauty and the low rents.

When Hayden arrived on Peaks he was a landscape painter, but he found that "the safety of the island" allowed him to relax, open up, and make a successful transition to abstraction. Last year, Hayden moved off the island after 13 years, feeling that he had absorbed what the place had to teach.

"I came to see the water as a moat," says Hayden, "and the whole move to the island as a spiritual journey."

All good art is a search for meaning. What artists find on Maine's island depends not only on what they are looking for — privacy, inspiration, natural beauty, society — but also on the sensibility of the artist and the character of the island. Perhaps the most privileged island experience is reserved for those who inhabit private islands or own one of their own.

David Rowe, one of New York's rising stars, for example, has summered on Cushing's Island in Casco Bay since boyhood, his intellectual abstractions drawing some of their clarity from the quality of Maine light.

And Andrew Wyeth, while more closely associated with the mainland town of Cushing, has

Monhegan Headland (detail), Reuben Tam, 1968

The Dock, Fairfield Porter, 1974–75

created paintings of sublime beauty and eeriness on Allen Island, the historic island he owns with his wife, Betsy.

Perhaps the most lovingly celebrated of Maine's private islands, however, is Great Spruce Head in the bright heart of Penobscot Bay. The summer retreat of the Porter family of Winnetka, Illinois, Great Spruce Head was painted by Fairfield Porter and photographed by his brother Eliot throughout much of the mid-20th century.

A painterly realist in love with the French Intimists, Fairfield Porter painted the family summer life of Great Spruce Head in sure, sunny tones that make of the island a kind of saltwater Eden. Commenting on one of his island paintings in Alan Gussow's *A Sense of Place,* Porter observed that the most important aspect of a place is "the quality of love."

"Love," wrote Fairfield Porter, "means paying very close attention to something, and you can only pay close attention to something because you can't help doing so."

An island, by its very nature, focuses one's attention, limits the cast world to a manageable few acres where close inspection and introspection is rewarded. Eliot Porter, whose 1966 appreciation of Great Spruce Head, *Summer Island,* remains one of the great books on Maine island experience, credited the island with making him both a naturalist and a photographer — an artist.

"So, it was that we began to live by lunar time," wrote Porter of his boyhood explorations of Great Spruce Head's shores. "A deep feeling for nature began to grow in me, a feeling that was to affect the whole future course of my life."

Talk to any artist who has worked on a Maine island, view the evidence of their art, and it becomes clear that island life, island time, is a powerful tonic, a transformational experience that re-charges creative energies and uplifts the human soul. For what time spent on a Maine island ultimately affords the mortal mind and eye is rare comprehension of wholeness.

Monhegan artist Reuben Tam, writing in his journal on June 2, 1965, expressed this perception more poetically. And it is well to end as one begins: "And today," wrote Tam, "a day of cold, of drizzle, of the sea fog blowing in and the cold wind hurling the north onto the land, the utter desolation of time, and of gray air and gray light, the foghorn sounding over the island — the aloneness of living itself — these, the sea, and the high cliffs on the other side of the island, and the presence of spruce trees denoting the north, make me realize, if only sporadically, that I have found much of what I have searched for here on Monhegan."

1993

Monhegan's Mold Cast Kent

Elliot Stanley

When Rockwell Kent stepped off the Monhegan mail boat from Boothbay Harbor in June 1905, he had completed two years of architectural studies from Columbia University, and an additional two years of art classes under New York artists William Merritt Chase and Robert Henri. He had also developed an intellectual fascination with socialism, as a rebellion against his upper-middle-class origins in Tarrytown.

Kent's studies in New York focused on realism in art, which Henri taught to his students — who included not only Kent but also Edward Hopper, George Bellows and George Luks. Collectively they helped provide the foundations of what came to be known as The Modern School. Hopper and Bellows stuck largely with the urban scene while Kent went first to Monhegan, then to interior New England, Newfoundland, Alaska, Tierra del Fuego, Ireland, and Greenland, before finally coming home to the Adirondack region of New York. After leaving the city for Monhegan in 1905, Kent never again used urban settings in any of his paintings.

As Yeats once advised Synge to go to the Aran Islands to rediscover the traditions and language of old Ireland, so Henri sent the young Kent to Monhegan to find the elemental in both the natural and human landscape. Henri had painted along the Maine coast, including Monhegan, and Kent's early works there show the influence of bold color and strongly contrasting forms typical of Henri's mature work.

Above: *Late Afternoon,* Rockwell Kent, 1906
Rocks, Monhegan, Rockwell Kent, 1907

Winter, Monhegan Island from 1907, considered one of Kent's major works, was given to Henri in acknowledgment of Kent's artistic debt to his teacher. It now is the sole example of Kent's painting on view at the Metropolitan Museum of Art.

A 1928 essay Kent wrote reveals the island's impact in developing his artistic and social consciousness:

"I wanted more than anything to live on the ocean; so I went to Monhegan Island. Because I had never done any work with my hands I was most impressed by the strength and potential power of people who did work with their hands. Seeing fishermen at work in an element that was terrible to me, I felt my own inferiority and the necessity to restore my self-respect by learning how to work."

In the painting *Toilers of the Sea* (1907) we feel the straining of muscle against the sea in the daily work of the Monhegan lobstermen. His intellectual interest in socialist theory was bolstered by his growing awareness of the harsh reality of life faced by working people; he identified his artistic craft with the work of the common man, and throughout his life he joined unions and worked with his hands at all stages of the engraving or printing of his own works.

Kent's first Monhegan period lasted off and on from 1905 to 1910. During those five years he produced about 120 oil paintings or preliminary studies for them — one of his most prolific periods of work. As a group, the Monhegan paintings won immediate critical recognition in New York (despite little sales success before 1919), and are increasingly regarded as comparable in stature to Winslow Homer's seascapes. In those early years he also designed and constructed four buildings on the island — his only known architectural work in Maine. One of these houses, the Sara Kent cottage (for his mother's use) is now owned by artist Jamie Wyeth.

Kent's second and last Monhegan period consisted of summers there between 1947 and 1953. Again he was moved to paint, but the 40 intervening years of highly publicized adventures, world renown as artist and book illustrator, and political controversy had brought him back to Monhegan seeking refuge from the world, rather than a jumping-off point for more excitement. Kent completed 36 paintings during his late 60s and early 70s.

If the artistic fire had cooled somewhat by the time Kent returned to Monhegan, the social conscience and political activism that had burned brightly since youth had, if anything, intensified in

Toilers of the Sea, Rockwell Kent, 1907

Winter, Monhegan Island, Rockwell Kent, 1907

his later years. His ardent socialism and membership in the Communist-front organizations, while hardly unusual among American intelligentsia before World War II, exposed him to increasing criticism in the Cold War era, including his celebrated confrontation with Senator Joseph McCarthy in 1953.

There were also hurts associated with his high profile as a famous and politically outspoken artist. Of these, none was more poignant than the 1953 decision by the Board of the Farnsworth Museum at Rockland to cancel plans to build a special Rockwell Kent wing; in return, Kent was prepared to give the museum his own personal collection of more than 75 canvases and hundreds of graphics, drawings, and manuscripts. This would have been the largest collection of his works under one roof in the world. When Kent heard that Boston banking interests on the board vetoed the plan based on his politics, he was outraged. The years of Rockwell Kent's association with

Monhegan and with Maine thus came to an abrupt end in 1953. The security of his "unworldly" refuge had been violated by its own provincialism — even on Monhegan he felt the sting of suspicion and rumor.

In 1961 he gave to the people of the Soviet Union the vast art collection he had hoped to have at the Farnsworth; they in turn accorded him the Lenin Peace Prize in 1967, the Russian "Nobel."

Shortly before his 89th birthday in 1971, Kent died of natural causes in Au Sable Forks. He had cared deeply and passionately about art and about people, and he had worked hard over a long life both for his art and for the advancement of his political ideas. His art, a powerful instrument, and his radical thinking made him a threat, and a fearful country ostracized him. But his association with Maine and Monhegan has undeniably enriched all of us, on both sides of the cold Atlantic.

1988

Southern Island: Father and Son

Christopher Crosman

Southern Island and its lighthouse, signaling the entrance to Tenants Harbor, have been the subject of dozens of remarkable paintings over a period of the last 15 years by Andrew Wyeth and his son Jamie. Not much more than a dot on many nautical charts, Southern Island is approximately 22 acres of grass, rock and a few windswept spruce. It is only a few minutes from Tenants Harbor by boat, but the lighthouse itself, located on the opposite side of the island, looks out to a sweeping view of the open Atlantic with only the faintest trace of Metinic Island in the far distance. The fetch, they say, is Spain.

The lighthouse, dating to 1857, was extinguished in 1934 and restored in the late 1970s by Betsy James Wyeth. It became Andrew and Betsy's private retreat until 1990, when they moved to another island farther out to sea off Port Clyde. Eventually, Jamie Wyeth moved his principal studio and residence to Southern Island from Monhegan, in part to find more privacy for his work.

While the Wyeths, father and son, rarely stray far from home for their subject matter, their Southern Island paintings are exceptional for their sharp, intense and prolonged focus on this small patch of land.

Light Station (detail), Jamie Wyeth, 1992
Flying High, Andrew Wyeth, 1987

Just as the prisms of a lighthouse lens strengthen and concentrate a beam of light, so, too, the paintings by both artists illuminate this island with flashing clarity and emotional precision. With its connotations of searching and illumination, Southern Island lighthouse is a potent metaphor for art of both Wyeths.

Comparisons between Andrew and Jamie Wyeth's paintings on Southern Island would appear inevitable and obvious. However, as conveyed through their antipodean visions, Southern somehow becomes very different, infinitely mysterious and variable, as if seen through the opposite end of the same telescope....

Light and dark, near and far, solitude and presence, past and present, timelessness and change — the trope of the artist watching and seeing are all contained in Andrew's spare, concentrated watercolors of Southern Island. In terms of abstraction, Andrew Wyeth delights in the infinite variations of white and in the textures of wall and fabric. *U.S. Navy* is a drybrush portrait of an African-American enlisted man with an authentic War of 1812 officer's tunic hanging on the wall behind him. Again there is a strong sense of presence through absence, a kind of relaxed informality against a harsh and turbulent seafaring past that held no glory for African-Americans in the early years of our country's history. He uses the deep blue of the jacket and the sailor's dark flesh tones to create a kind of visual ping pong between the figure and background, thereby heightening the psychological tension and ironic juxtaposition in the otherwise cool and emotionless portrait. The officer's coat, perhaps, has deep personal meanings as well: the artist's father, N. C. Wyeth, was working on illustrations for an edition of the Horatio Hornblower series when he was killed in a tragic train-car accident in 1945, just as Andrew was beginning to claim his own career as an artist. The cover illustration for one of the Hornblower books, with a figure wearing a similar jacket, was completed by Andrew shortly after his father's death in what must have been a wrenching posthumous collaboration. It would be the last illustration commission that Andrew Wyeth ever accepted.

The officer's jacket reappears in *Dr. Syn*, among Andrew's most unexpected and startling works. The title refers to a favorite Wyeth film starring George Arliss, about a pirate who becomes a minister. It is also a self-portrait; the artist even had an X-ray of his skull taken for the

Meteor Shower, Jamie Wyeth, 1993

Dr. Syn, Andrew Wyeth, 1981

figure. The setting, which resembles the gun deck of an early 19th-century English or American warship, is, in fact, the interior of the fog bell tower on Southern Island, as remodeled by Betsy Wyeth to serve as a studio for her husband. Andrew, in turn, presented *Dr. Syn* to Betsy for her birthday; she promptly hung it in the living room at Southern Island. While such playfulness is a Wyeth family tradition, the painting gains poignancy in light of

the artist's reference to this literal mantle passed from father, N. C. (the jacket was originally owned by N. C.'s teacher, Howard Pyle), to son, Andrew (who has since passed it to Jamie, as seen in several recent works). Wyeth friend and biographer Richard Meryman recently observed, "[Andrew] Wyeth, using many objects and people, has continued to paint his father. N. C. remains a central, perhaps hourly, presence in his life. . ." "My father

Squall, Andrew Wyeth, 1986

is still alive," he says, "I feel my father all around."

There is, of course, a great deal more to Wyeth's art than this complex, emotional subtext of loss, coupled with a natural but no less terrible sense of freedom from his father's powerful influence. Still, absence seems to be a recurring theme in many of the Southern Island paintings, most notably in *Squall* and in *Battle Ensign.* The latter, in Wyeth's mind, came to be about his old friend, Walter Anderson, who died while Wyeth was working on the painting. The ensign is seen reversed and tattered, a symbol of the sea and blood and fighting. Although some viewers find the gritty toughness of the image unsettling, Wyeth simply responds, "I liked it because it was wrong."

A different Southern Island Light is the subject of Jamie Wyeth's recent paintings. But, as noted by James Duff, director of the Brandywine River Museum, many of the same qualities of human endurance, the mutability of nature, solitude and quiet, wit and humor, and the universal significance found in the local and ordinary are also present in the younger Wyeth's paintings, even as his approach to art is profoundly his own.

His island is a place where reality collides with a world of myth and magic, strangeness and wonder. The narrow gulf separating the two is suggested by *Meteor Shower,* one of the artist's most enigmatic paintings. A scarecrow with a strange, bird-like, leather mask and the War of 1812 jacket

(the same jacket in *Dr. Syn* and *U.S. Navy*) is silhouetted against the night sky and the lights of Tenants Harbor a short distance across the water. The creature seems to cock its head, as if it hears nature's fireworks in the sky overhead. A portrait of the artist and his uncanny gifts and strange craft? His isolation and self-imposed exile from those 'normal' folks across the harbor? Like many of Jamie Wyeth's works, there is a mournful note of deep melancholy interposed with humor and self-deprecation.

Magic is implied in several paintings of pumpkins. *Lighthouse Pumpkin* and *New England Pie Pumpkins* recall Halloween, a favorite Wyeth family holiday, from the times when N. C. and the family would put on costumes and makeup to bring to life the stories and myths he illustrated. So, too, is the memory of N. C., the grandfather Jamie never knew who was killed shortly before Halloween, invoked with affection and humor in the raucous gathering of pie pumpkins before a door slightly ajar.

Several years ago when the "storm of the century" was predicted to hit the Maine coast, Jamie, and his wife Phyllis, packed their car and hurried back from their Delaware home to Southern Island (despite warnings that coastal communities might be evacuated). The storm was less ferocious than originally predicted, but Jamie found that the light tower "screamed like a dozen Metroliners" as the wind raced through the lantern flues. There is a streak of wildness and testing limits, of life under extreme conditions, in Jamie's art. Therein, not coincidentally, lies the essence and history of lighthouses in Maine.

Increasingly Jamie spends long periods on the island throughout the year. He spent weeks waiting for fog, the subject of *Light Station*, and since he paints directly, he could only work for minutes at a time on the painting when the fog finally rolled in, because the palpable dampness would not allow paint to adhere to canvas. The final painting is, however, fiction: neither the lamp nor the figure (Orca Bates, who has modeled for numerous other paintings) were present in reality. The lamp—here represented by a zig-zag flash of yellow that quite literally gives this work an electric charge—has long since been removed; Orca was added from memory. Not since Whistler's *Battersea Bridge* has anyone captured in oils so convincingly the tangible stuff of fog, its caressing, contour-eroding thickness and tonal complexity.

Wyeth ultimately creates an emotionally powerful image reflecting the inner lives of his protagonists and through them a more universal statement about reality and illusion, human fragility and permanence. Variations on this theme can be found in several ravishing paintings of flowers and plants ranging from the potted vegetable garden in *Lighthouse Garden* to the brilliant irises near the light tower in *Bees at Sea* and *Iris at Sea*. All refer indirectly to the improbability of survival, much less beauty, found on this windswept, rocky and sea-battered island. In *Iris at Sea*, the fragile, sun-bleached flower struggles tenaciously to lift itself against the looming lighthouse, which alone endures beyond the fleeting Maine summer.

If the iris is a metaphor for survival and tenacity, then Jamie's recent portrait of Phyllis, *Southern Light* (1994), is one of his most direct and compelling paintings to date. Framed by the sturdy, ornamental detailing of the doorway to the bell tower and suffused with a golden light, Phyllis stands before us with quiet dignity, erect and proud in the manner of countless images in literature and art of New England women and the sea. It is a miraculous, loving portrait of beauty, inner strength and endurance through grace. If the presence of N. C. Wyeth lingers in both artists' depictions of the lighthouse, the indomitable spirit of Phyllis Wyeth is embedded in the very granite foundations of Jamie's art.

Southern Light, Jamie Wyeth, 1994

By the Light of the Moon, Andrew Wyeth, 1987

Battle Ensign, Andrew Wyeth, 1987

Acutely aware of the long tradition of light-houses and islands in Maine art, both Wyeths uniquely evoke the earlier, Emersonian ideal of spiritual communion through the smallest details of nature. The Southern Island paintings must also be viewed in the context of their extraordinary sense of place, although each sees this place quite differently, in terms of emotional tone and empha-sis as well as style and technique. Moreover, theirs is a modernist sensibility of self-awareness, of a kind of fluid deliquescence between form and sub-ject, and precise, charged emotional intensity. Indeed, the sheer number and variety of views of this particular island suggest parallels with a con-temporary "serial" esthetic found in DeKooning's "women" series and Robert Motherwell's "Elegies to the Spanish Republic," or even the multiple soup cans of Andy Warhol (whose portrait Jamie Wyeth has painted). For the Wyeths, repetition, or, rather, continuity and concentration of vision, is like the steady, insistent gonging of the bell buoy or the beaming intensity and sharpness of the tower light. It is, then, a matter of a particular lighthouse becoming a place of continuity with the American landscape tradition, of continuing self-discovery and connectedness to family, of revela-tion and emotional rescue, as well as, perhaps, a cri de coeur for contemporary life's loss of insight

Bees at Sea—Study # 1, Jamie Wyeth, 1994

— possibly the truest metaphor for the long-dark
Southern Island Light, now illuminating realms of
the spirit and imagination, through the paintings of
Andrew and Jamie Wyeth.

1995

Writing With Light

Philip Booth

PHOTOGRAPHS BY JEFF DWORSKY

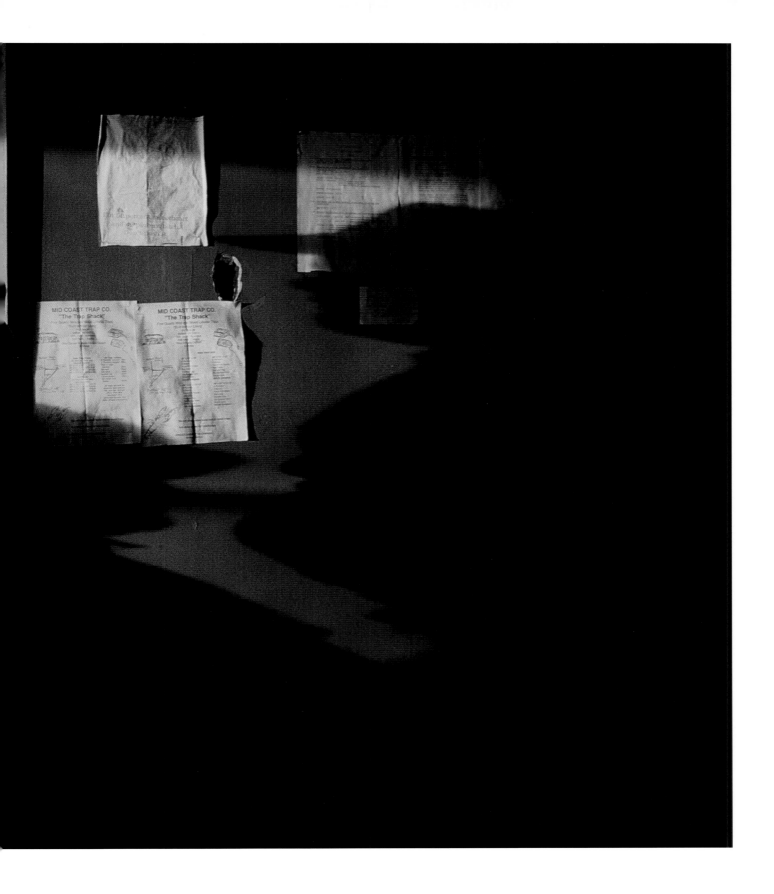

Photography is, literally, writing with light….As his photographs clearly prove, Jeff Dworsky's eye is used to seeing through a viewfinder. Seeing, not merely looking. Whether he's carrying a camera or not, his eye observes, his mind fills with insight, his heart responds — here to available light of humanity itself — with an acuity that no auto focus or exposure meter can measure. With camera or without, he sees potential photographs; when he has camera in hand he has already readied himself to record what Henri Cartier-Bresson called "the decisive moment," that fraction-of-a-

His photographs show, tellingly, how faces feel.

second when what is vital in the viewfinder is fully self-illuminating.

No one who hasn't weathered a whole year on an offshore Maine island could possibly have made these photographs. No one who hasn't lived long on islands can come close to knowing the terms of island life, especially the isolation that December darkness brings home, and the weight and torque and blindness of midwinter storms. First as a clam digger, later lobstering with an outboard, going sternman on a scalloper, and then fishing the first boat he built (under Arno Day's sharp eye), Jeff Dworsky has worked the tides in all seasons, and has earned his own local knowledge of ledges not marked on any chart, and weathers NOAA never predicted. For some 15 years he has brought his catch home to Deer Isle, York Island and now Isle au Haut.

From just such various islands, all within seven miles of each other, Jeff Dworsky's camera work demonstrably brings home to mainlanders, no matter how far inland, the harsh immediacies of everyday island life. His lens shows what the instant of his shutter's opening tells: the intense aloneness that, paradoxically, makes for a commu-

nity bound — not without twisting and tugging — by necessity. Only an island community grown to tell the hard and redeeming truths of island life; only an islander grown all-but-native could know how to open his camera in non-invasive ways to the faces and lives of the people he lives and works with, and cares about deeply.

Even in summer, islanders are, by self-definition, a breed apart. And proud of it. Given what passes for civilization on the mainland, it's small wonder that the wife of a "well-spoken" lobsterman "met by chance on Swan's Island," once softly told a California poet, "What I like more than anything is to visit other islands." Islanders value their island values, their mostly unspoken codes, with a gentleness that is equaled only by the violence of island fishermen settling the constant small wars of their territorial rights to drag for scallops or set lobster traps where their fathers fished before them. Whether islander by birth, or islander by being a loner or in love, any islander instinctively knows that she or he is the equal of any other islander; and more-than-equal to mainlanders, especially those from away who flaunt dollars or try to talk as if they could speak the rhythms and words of island speech. Maine islanders are most of all equal not only to each other, but to the inescapable fact of being islanders. No matter what modern worldliness radios and TV dishes bring to an island, its year-round inhabitants incomparably realize what it means to root one's life on a low mountaintop shaped, constricted and surrounded by the sea. Lifelong islanders, if faced with such dying as still gives them marginal choice, more often than not refuse to be moved to hospitals on the mainland: they know without doubt in their inmost selves that they want to die where they were born to live.

Living surrounded by tides, and working those tides for a living, is the life Jeff Dworsky lives and photographs. His photographs are in many ways the color equivalents of the black-and-whites John deVisser made of fishermen and fisher-families on the South Coast of Newfoundland 30 years ago, images that were altogether co-equal with the text of Farley Mowatt's *This Rock Within the Sea: A Heritage Lost*. Like deVisser, Dworsky photographs the human resilience of people who inhabit a rock within the sea; but Dworsky's viewpoint is more surely his own. Mowatt painfully writes of discovering a "heritage lost" in the process of his writing and of deVisser's camera work. Jeff Dworsky similarly documents a heritage, but precisely because he is part of that heritage, his photographs argue the more strongly for its survival. They particularly challenge the expensive assumptions many of us have about living a life that appears to insulate us from the existential realities that islanders daily confront.

Jeff Dworsky's eye cares little for mere appearance. It cares duly for handsaw, hammer, potwarp, engine oil, starting fluid and boat paint. It cares passionately for human beings who know they live on one of this island world's smaller islands, islands where life's daily terms are daily visible. Notably visible in all but a few of these photographs is the window through which islanders forever look out, in whatever weather, to check the weather and, if it is visible, to see the sea. In winter's hardest weathers they barely leave the house, much less the island. With so much beyond all control beyond them, generation after generation of islanders know that if they're to winter out, to

Only an islander grown all-but-native could know how to open his camera in non-invasive ways to the faces and lives of the people he lives and works with, and cares about deeply.

No one who hasn't lived long on islands can come close to knowing the terms of island life.

make it over the old March hill, they must snug down in their kitchens, their general store, stay close to woodshed, workbench and woodstove. And, without choice, live close to each other.

In habiting such necessity, Jeff Dworsky's camera eye focuses most acutely on these basics of island life: work, love, joy, anguish, prayer and play; on how such basics are equally shared, whether in ways lonely or communal, by women, men, children, of age after age after age. His photographs show, tellingly, how faces feel. And how, in every aspect of island experience, hands work and express. How, in extreme human weathers, hands cover face.

Jeff Dworsky has, in an island context, made interior photographs which tell, as quietly as a Chekhov story might, how we humans survive in our most interior selves. To refamiliarize ourselves with the islanders of these photographs is to review isolated aspects of our own lives, to find anew the essential community to which we inescapably belong.

1992

NORTH HAVEN
In Memoriam: Robert Lowell

I can make out the rigging of a schooner
a mile off; I can count
the new cones on the spruce. It is so still
the pale bay wears a milky skin, the sky
no clouds, except for one long, carded, horse's tail.

The islands haven't shifted since last summer,
even if I like to pretend they have
— drifting, in a dreamy sort of way,
a little north, a little south or sidewise,
and that they're free within the blue frontiers of bay.

This month, our favorite one is full of flowers:
Buttercups, Red Clover, Purple Vetch,
Hawkweed still burning, Daisies pied, Eyebright,
the Fragrant Bedstraw's incandescent stars,
and more, returned, to paint the meadows with delight.

PETER RALSTON

The Goldfinches are back, or others like them,
and the White-throated Sparrow's five-note song,
pleading and pleading, brings tears to the eyes.
Nature repeats herself, or almost does:
repeat, repeat, repeat; revise, revise, revise.

Years ago, you told me it was here
(in 1932?) you first "discovered girls"
and learned to sail and learned to kiss.
You had "such fun,, you said, that classic summer.
("Fun" —it always seemed to leave you at a loss...)

You left North Haven, anchored in its rock,
afloat in mystic blue... And now—you've left
for good. You can't derange, or re-arrange,
your poems again. (But the Sparrows can their song.)
The words won't change again. Sad friend, you cannot change.

Elizabeth Bishop,
Island Journal 2003

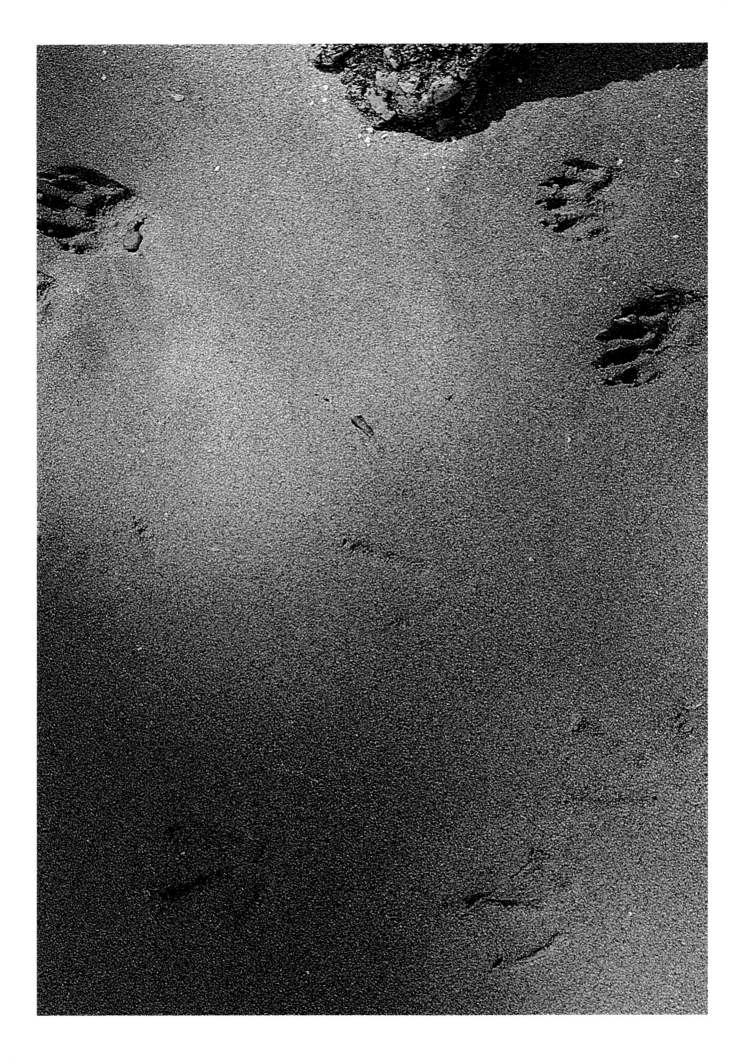

CHAPTER SEVEN

Birds & Beasts

ISLAND JOURNAL HAS BEEN A NATURAL HISTORY publication from its inception. Early issues are full of articles on auks, terns, gulls, geology, forests, extinction, evolution — the players, forces and processes, in short, in the drama that has always shaped islands. "Islands are sensitive ecosystems supporting populations of plants and animals predisposed to extinction," wrote Richard Podolsky in a 1988 essay on the subject. "It is crucial that islands everywhere be recognized and treated as the fragile and vulnerable ecosystems they are."

"Fragile" is not to be confused with "pristine," cautions the late ecologist Bill Drury in an article included here. As Drury and others have observed, Maine islands have been transformed more than once by the actions of human beings. "We now know that laissez-faire ecology, like laissez-faire economics, doesn't lead to balanced systems, it leads to monopolies," Drury writes after documenting the interactions of terns, gulls and people on islands in the Gulf of Maine.

Harry Thurston, whose 1993 article on the workings of the Bay of Fundy is included here at its original length, writes that by whatever name scientists choose to call themselves, "we all take it upon ourselves to understand, describe, and ultimately conserve communities of life — that is, ecosystems." Thurston's goal may be political (he opposed the construction of a tidal power project in the Annapolis Basin), but his description of the Bay of Fundy and its natural systems seems destined to outlive its original purpose and become a natural-history classic.

The best writers of natural history know that the observations of lay observers often contribute to the advancement of knowledge as much as the work of scientists. "Islanders know their places well," says naturalist and raven expert Berndt Heinrich, "and have been helpful resources aiding in other wildlife research. They are aware of populations of birds, the times of year certain species arrive and depart, and how they behave when there." In the study of natural history, the message is always that islanders have much to contribute.

David D. Platt

Peter Ralston (Opposite)
Above: *Urchin* (detail), Jamie Wyeth, 1999

A Stirring
of Ravens

Susan Hand Shetterly

In the shifting fog at sunrise, one can stand on Monhegan's cliffs and watch the long, dark perpendiculars appear and vanish and appear again. The water below them drums. Out of the fog a silent raven flies, dips before a cliff edge and is gone.

"I am not able to speak any good words for [ravens] as I cannot discover any beneficial habit save that of scavenger while they certainly are a menace to other birds and sheep," Ora Willis Knight fusses in *The Birds of Maine*, his 1908 volume. The science of the 19th and early 20th centuries is leaden with moral judgment — as if a wild animal should wish to cultivate "beneficial habits" and the blessing of a man's "good words."

In medieval Western Europe, the raven owned a reputation, along with the wolf, as denizen of the battlefields. The birds feasted on corpses; they plucked the eyes of the dead. No matter that men did the killing, the second act, the clean-up, horrified human witnesses.

Perhaps, as Barry Lopez suggests in *Lives of Wolves and Men*, the fabled rapacity of animals such as wolves and ravens held up a mirror men didn't prefer. And, Lopez continues, there is something more, something rooted in Western man's distrust of wildness. Wolves are not dogs. Ravens are not doves. Send a raven on a task, as Noah did the dove, and it takes off on its own. It doesn't report

Ravens in Winter, Jamie Wyeth, 1996

back. When Europeans settled New England, they brought — along with their enormous energy and hope — their worn antagonisms.

Edward Howe Forbush published his famous *Birds of Massachusetts and Other New England*

States in 1927, and discovered there were no breeding ravens left to report in his state. They had been extirpated, like the wolf. The birds' last New England strongholds, he wrote, were the Maine islands.

Today the raven is a popular bird. We recognize its honored place in the lives of native peoples of the Northwest, and we invest it with almost supernatural sagacity. Out on the Maine islands, where one can often see a raven lifting in

Offshore Raven (detail), Jamie Wyeth, 1996

the sea wind, and hear its voice — the calls sometimes soft, sometimes harsh, sometimes almost the same deep timbre as a buoy marker — it is still essentially unknown.

Studies of terns and puffins, peregrine migration and cormorant diets have taught us something about the value and richness of our island ecology. But we have overlooked a bird that probably affects life on a number of islands as much as any of these more celebrated species. How do ravens use these various juts of land off the Maine coast? Where do they feed and sleep and nest?

On a hillside in western Maine, not far from Mount Blue, Bernd Heinrich sits quietly on the mid-branches of a white pine, watching a band of ravens rend the open carcass of a moose. He has been watching ravens, and writing about them, for years, and he has little use for the historical reports of the birds. In his 1989 book, *Ravens in Winter*, he remarks, "More has probably been written about the raven than about any other bird ... but definite scientific studies are very few ... most of the literature consists of notes and anecdotes, and

many of the conclusions are false or misleading. Furthermore, much of our 'knowledge' is clouded (or illuminated?) by centuries-old myths . . . even now the raven is truly a bird of mystery." Heinrich may, at this point, know more about the mystery that is the raven than anyone else alive. He calls them the brains of the bird world, and has raised young ravens himself for his studies.

Born in Poland, Heinrich became a refugee, along with his parents and sisters, after World War II. His 1984 book, *In a Patch of Fireweed*, recounts the family's pilgrimage from a hut in a German forest to a farm in the hills of western Maine. It is the story of a keen, sensitive boy not bound in by war or by the strictures of a rural American town — but, rather, released to woodlands, to the secrets and revelations of the natural world. Now a professor of zoology at the University of Vermont, Heinrich returns to the hills where he grew up to make his own painstaking science. He studies the birds of the foothills and mountains of the interior.

The distances between islands, their inaccessibility, their often inhospitable isolation, kept them, in the years of white settlement, free from the relentless shooting and poisoning of wolves and ravens that took place on the mainland. Farmers did pasture sheep on some islands, and attempted to destroy ravens when they found them near. But they didn't eradicate them. The islands in the Gulf of Maine became safe havens for the species. This is why, Heinrich insists, it is so important to protect large buffers of wilderness. They often become refuges where remnant populations of wildlife are protected, and from whence they can radiate outward again when conditions are favorable.

"Despite relatively recent persecution," he writes, "the raven has been making a dramatic comeback in New England . . . ravens came to central and western Maine thirty years ago, apparently close on the heels of the invading coyote."

Ravens, despite their heavy, chisel-shaped beaks, require some other animal, such as a coyote, to open a carcass. Their beaks cannot penetrate the hide of a deer or a moose.

Ravens in Winter is about the journey of a question: Why do ravens flock and recruit each other to carcasses? If food is scarce, what purpose

Saltwater Ice, Jamie Wyeth, 1997

does it serve an individual, Heinrich asked, to share it? Ravens are generalists — their palate is broad. But, Heinrich writes, "they are nevertheless highly specialized carrion feeders."

He has found that a breeding pair of ravens defends a large territory at point sources: at their nest, at carcasses. But numbers of juveniles and other adult birds will invade a territory guarded by a mated pair to pillage the carcasses within it.

After five years of work, Heinrich concluded that raven recruiting behavior summons a group because trespassing is safer in numbers; he has also found that ravens at food sources play out the serious games of dominance and pair-bonding. They eat, they joust, they select their mates.

How might these behaviors differ on islands with long stretches of water in between? Sitting in his cabin near Hills Pond, Heinrich turns over the question in his mind. As he imagines transposing the behaviors he has learned in the foothills to the islands in the Gulf of Maine, he formulates other questions: Might ravens on islands find abundant winter food along the shorelines? Exactly how do

Ravens, Standing, Jamie Wyeth, 1996

ravens use the summer nesting islands and the seal-pupping ledges? Do ravens drift between islands and the mainland? How far will they fly out over water? Where do the young go after they fledge from island nests? Do island ravens without mates eventually wander inland? And what opens the carcasses of dead animals for the birds on islands without coyotes?

People who live on islands would be aware of the nesting and the roosting because ravens are noisy birds. Islanders know their places well, and have been helpful resources aiding in other wildlife research. They are aware of populations of birds, the times of year certain species arrive and depart, and how they behave when there.

Each island, with its distinct and separate set of variables, is an excellent place for controlled observations. Heinrich would begin his study with the distance from the mainland; he would investigate possible food sources such as deer, seals, mollusks, berries, eggs and chicks; nesting locations on cliffs or in stands of spruce; roosting locations and the presence or absence of the birds at different seasons. There are the questions of territory and recruitment. Might a raven pair claim islands as territory? Would it be possible, over long stretches of water, for other ravens to assemble and invade as they do inland?

Ravens make a variety of sounds that amount to language. Heinrich has interpreted the screams, quorks, and dry rattles — testing what he thinks they communicate against the particular occasions in which they are uttered. Might ravens along the coast and islands make sounds that stand for occurrences that are exclusively maritime? If so, we might learn that the ravens of Maine have evolved — like Maine's own people — various dialects.

Heinrich can imagine what the cliffs of Monhegan might reveal about the birds he has had the privilege of studying elsewhere for so many years. Those cliffs are far away. Yet he believes it is time to begin to know the ravens of the island places that have been, for so long, their best safe havens.

The islands of Maine are like petri dishes into which one arranges specific trees and cliffs and outcrops, rocky shore or beach, a seal-pupping ledge, a deer herd, herring gulls, eagles, an osprey nest. Drop in the raven. Watch what it stirs up.

1996

RAVEN'S GORE

The steep stone beach is hot,
Where a yellow wind stirs eddies against
The skin and bones of Brimstone Island.
We have steamed all morning
To be here among the gulls
That scream at us from above.

I choose to bring the children here
To celebrate this Father's Day,
Knowing we will not find you here;
Knowing we can only find you here,
Where we've been so many times before
At this entrance to the island of the dead.

Visits to these cliffs years past always
Felt so large, and time so immense;
Expanding out over the whole ocean
Like inner vision from an immortal hawk.
Days when little boys climbed the hills,
Heads poking from brown grasses
To gain some purchase on this place
Where one world meets the next.

Two decades or more ago
Here before I even knew of you,
Edging around an outcrop,
High above the breaking surf
I came face to face with a raven
Hunched on a narrow ledge
Staring balefully out to sea.
Suspended and motionless,
Neither of us dare breathe
Until the spell is broken
And the raven flings itself
Out into the void,
A broken wing trails uselessly
As it flutters to the sea,
And paddles outward bound
With its one wing waving.

Instantly the gulls are alerted, seeing
Something living, but only partly living.
They wheel and scream:
Raven down! Raven down!
And swoop on the sodden creature
To square accounts for once
With this ancient bird of death.
On its back, it rises up again and again
To greet each new tormenter,
In the agony of its impending doom,
Meeting each feint with an eerie cry
Until a clever pair time their assault
And break its neck from behind.

As we stand here with your ashes,
Three ravens grok about in a lone spruce
And then two more wing in from the west.
Five ravens now over the five of us.
Oh how your numbers have grown!
Here now with you gone
But forty-nine days ago.

I wish there were some way to make peace
Amid these warring fates
That contend for this most sacred space.
Is it your restless spirit or mine
That unleashes these winged furies
Where the sea just pounds the shore?
But, oh! the sheen this endless rote gives
To brimstones rolled and tumbled
And piled soul-deep on this beach,
Polished parts of lifeless creation,
Now in each chamber of stone
Where I still hear your defiant call
I will carry in the inner ear
With my love to the grave.

Philip Conkling, *2000*

Untitled Study, Jamie Wyeth, 1996

The growth of the herring gull population and southward extension of their breeding range continued during the 1950s, to Connecticut, New Jersey and Maryland. The first herring gull colonists reached North Carolina about 1960, and the colonies south of New York continued to grow in the 1970s and 1980s....

Once an area is colonized only a relatively small number of gulleries is needed to produce the young birds that keep the population going. Reproduction on other islands is not important.

This means two things: first, that eliminating gulls from a number of outer islands where terns might nest will have no effect on "reducing the gull population." Second, that our concern for terns must be focused on those islands where parents produce enough young to export. These colonies will be able to maintain the population, but keeping terns nesting on islands where parents fail, provides a population sink. To manage a tern population we need to make sure when we attract birds to a nesting island that they breed successfully....

The population of gulls nesting in Maine grew only slowly during the 1950s, and some can argue that [an egg-spraying program instituted in the 1930s] was working. The population in Maine has stayed "conservative" since the program was ended, and it seems likely that the effect is the same as that of the gulls moving from outer islands to the inner islands in the 1930s and 1940s.

Between 1967 and 1973, we worked with Biological Services agents John Peterson and Frank Gramlich of the Fish and Wildlife Service to conduct control experiments at islands where gulls were encroaching on terns.

The exercise taught us that as we removed the nesting birds a number of gulls, especially immatures, replaced them. Killing gulls on tern islands was not successful because we had to "bleed down" the surplus of young gulls interested in, but excluded from, breeding.

Some people have recommended that the entire gull population or a regional segment be reduced. The reality of the reproductive potential of gulls and the fluidity of the populations suggest that this is a much larger undertaking than its proponents realize. It is a project that would require

Arctic tern

PETER RALSTON (2)

the cooperation of operators all over most of the East Coast herring gull range, and it is unlikely that such cooperation would be given. If the numbers of gulls were seriously reduced, they would probably first become scarce in remote places, because most gulls are attracted to food around towns and cities. It is more reasonable and effective to remove the attractions where gulls gather and become pests.

The problem of gulls is important in places where forces strongly attract a relatively small number of gulls. That is what we are dealing with at vulnerable tern colonies. But reducing the problems to a small number of areas and a minimum number of gulls is not good enough for some committed idealists.

Many people with whom I must debate killing gulls prefer to deny that any need exists. Some argue against killing on humanitarian grounds. I agree; killing is not a pretty or enjoyable activity. But gulls eating living tern chicks is not a pretty sight, nor is the sight of a herring gull pecking at the bleeding head of a living laughing gull that was just recently incubating its eggs.

I think that the philosophical question of killing one species to favor another was made and accepted by those early agriculturists who pulled up plants that inhibited the growth of their crops — they weeded the garden. A biologist can argue that it is precious and self-serving to make a philosophical separation of plants from animals, or "lower animals" from vertebrates, or us. Why should fish be placed "beyond the pale?" When we find vulnerable species that we think are important, and want to encourage them, we may have to seed the garden.

As to the philosophical issue of nature's order and "playing God," we now know that laissez-faire ecology, like laissez-faire economics, doesn't lead to balanced systems, it leads to monopolies. Some species will take over and assert an order favorable to them, whether white spruce or starfish or herring gulls. Unless we believe that there is a natural order established at the creation, we should acknowledge that when we won't play God, someone else will.

1989

The Life and
Tides of Fundy

Harry Thurston

To the uninitiated the muddy waters of the inner reaches of the Bay of Fundy might look like a watery desert. Closer inspection proves they are far from barren. Several years ago, I stood on the high marsh, watching as the brown tide inexorably transformed the mud-walled, dry-bottomed Allen Creek into a navigable channel. Marven Snowden, of Wood Point, New Brunswick, quickly powered his boat into the open waters of the Cumberland Basin where four generations of his family had fished for American shad, the largest, and most succulent, member of the herring family.

Marven had seen many changes in the fisheries of the upper Bay of Fundy. His grandfather fished during the heyday of the shad fishery at the turn of the century. Then, hundreds of thousands of pounds were salted in barrels for export to the Eastern Seaboard. During his father's time, the fishery had collapsed suddenly. In recent years, Marven had seen the shad making a comeback.

The burly fisherman steered toward the Nova Scotia side of the basin. He knew that shad move with the strongest currents. "You try to get in the strongest stream and work toward the slack

water," he explained as his son and another crew member set the two kilometers of gill net.

We cut power and drifted on the tide. Marven freely shared his local wisdom. There are three distinct runs of shad in the upper Bay, he told me; and fish come here to feed, not spawn.

In recent times, science has often ignored local data gatherers — the fishermen. Fortunately, Dr. Michael Dadswell, then a federal fisheries research scientist, stationed in St. Andrew's, New Brunswick, was willing to listen to Marven Snowden. Though he says that his first meeting with Marven in 1978 was a case of serendipity, Dadswell deserves credit for piecing together the information into a workable hypothesis.

Science sometimes advances in leaps, like salmon or poetry. Dadswell had a hunch that the

PETER RALSTON

Herring weirs, Grand Manan Island

increase in Fundy shad might be related to the restoration of major shad spawning rivers in the United States, such as the Susquehanna, Delaware, and Hudson. As anadromous fish, shad spawn in fresh water and then return to the salt water. It was a long standing mystery in fishery science, however, just where the shad went after returning to the sea.

Dadswell asked Marven to help him tag shad in the Bay of Fundy. The next spring the fluorescent dorsal fin tags began returning with southern postmarks. Eventually, he received tags from every river with a known spawning shad population, from Florida north to Labrador. He now believes that every shad comes to the Bay of Fundy at least once during its life history.

Other species return annually to the rich feeding grounds spread out by the ebb and flow of

Fundy's tides. No migration is more spectacular than that of the sandpipers, or "peeps," as Fundy denizens affectionately call them. The peeps flock to special places along the upper Fundy shoreline to feed and, it seems, to perform their aerial ballet.

Hundreds, thousands, tens of thousands of sandpipers spiral from the mudflats like snow devils, then string out in sinuous banners of flight. The play of light on their dark backs and buff breasts, as the flock banks in perfect synchrony, is designed to foil a raptor's strike. But to the appreciative observer, the birds' flight seems nothing less than joyful expression, like a musical chord or brushstroke.

Much of the sandpipers' time, however, is spent on the flats in more pedestrian fashion, doggedly bobbing up and down in pursuit of the "mud shrimp." These fatty, translucent morsels are tucked into the fine tenements of mud in astronomical numbers — 20,000 to 60,000 per square meter. On this side of the Atlantic, they are found only in the Bay of Fundy and Gulf of Maine, and then only in great enough numbers at several sites to attract major shorebird flocks.

One of these mud shrimp hotspots is Mary's Point, New Brunswick. From a sandpiper's view, it has everything: the fertile muds of Ha Ha Bay on one side, the salt marshes of Shepody on another, and in between, a crescent sandpit for roosting.

Sandpipers arrive here unerringly, on or about July 18, from their Arctic breeding grounds. For them, it is a fuel stop, or as one ornithologist remarked, "a fat station." In two weeks, they will double their weight, becoming winged butterballs. That envelope of fat will carry them on a three-day, nonstop flight over the Atlantic to their wintering grounds on the north coast of South America.

It has been known since Audubon's time that the inner Bay of Fundy is an important staging area for shorebirds. But it was not until the late 1970s that biologists came to appreciate that it is the most important shorebird site in eastern North America, annually hosting some 1.5 million shorebirds of 34 species. By far the most numerous are the semipalmated sandpipers. In a given year, one-half to three-quarters of the eastern North American population stops to feed on the intertidal offerings of Fundy's mudflats.

There to greet them for the last two decades has been one of the Maritimes' extraordinary naturalists, Mary Majka. Mary's restored farmhouse

and beachside cottage stand vigil over Mary's Point and its migrants. "We know from history that many species have been extinguished because they couldn't be as flexible as human beings," Mary once observed as we sat on the beach next to a roost of 30,000 sandpipers. "And I think that certain species are very much more dependent on special environments, and those birds definitely are. They certainly cannot survive without the Bay, its tides, and its beautiful mud."

The truth of Mary's words received official sanction in 1987 when Mary's Point was dedicated as the first Western Hemispheric Shorebird Reserve in Canada — a critical link in the migratory chain.

Naturalists, conservationists, ecologists, environmentalists. By whatever name we choose to be called, all take it upon ourselves to understand, describe, and ultimately conserve communities of life — that is, ecosystems. In my experience, there is another kind of environmentally friendly individual who does not necessarily articulate "green" principles, object lesson in how to merge economy and ecology.

Let me introduce one such person.

In the ice-free months, April to October, Clayton Eagles of Five Islands, Nova Scotia, can be found on the intertidal prairie exposed at low tide in the Minas Basin; one of 200 Breughel-like figures bent at the waist, arms outstretched in the shape of a human tripod.

In the 1940s, when Clayton started clamming, diggers took only the larger clams because the smaller ones were uneconomical to shuck. Clayton

STEPHEN HOMER (2)

Sandpipers or 'peeps' on migration

can't break himself of the habit. On every dig, he still plucks out only the choice, mature clams and leaves the rest. This traditional conservation technique seems to have been abandoned by the new generation of clammers forced onto the flats by hard times in the 1980s. This unchecked exploitation of the resource haunts Clayton.

"They think I'm crazy, I don't like diggin' everything. I don't know, clams got to have somewhere to start," he told me on a discouragingly poor digging day. "Years ago, there used to be breeding beds, we'd call them. We never used to dig them at all. They started diggin' them out. I think it made a difference. I just can't see it. You dig everything out, what's going to be left to grow?"

In Fundy, the great tides dominate all living things — from the microscopic glass house of the benthic diatom to the 70-foot leviathan, the fin whale. Survival here depends on the ability to adapt to tidal range and rhythm. I have found that the principle can hold as true for people as it does for plants and animals that live by and under the Bay's waters. And no person I met along the Bayshore lived more in tune with the tides than Five Islands' weir man Gerald Lewis.

For 40 years, he drove his horse and wagon over the bottom of the sea, his movements mirror-ing the ebb and flow of his tidally ruled environment. His V-shaped, mile-and-a-half-long brush weir was a giant fish basket woven of spruce boughs, saplings, and twine. It gathered passively, sometimes only enough flounder for table fare, other times enough herring and shad for salting, and occasionally a rare catch, such as a tuna, which Gerald cut into steaks for his neighbors. The weir was a permeable dam that was moderate in its catching power. And it was impermanent, as each winter ice carried the 1,500 hand-driven weir stakes to sea. How unlike a tidal dam, I thought. It would be permanent, and immoderate both in scale and in its impact on the environment: disturbing clam beds, burying mud shrimp, and chopping shad to bits.

I can only hope that when the dream of tidal power undergoes its inevitable revival the lessons of the last 15 years are not forgotten: the true power of the tides is in their ability to do biological work. Said another way, tidal life is tidal power. It took this native-born Fundy boy — adrift among tide pools, marsh hay, and mudflats — three decades to express that simple thought.

1993

Observation blind, Puffin Project, Eastern Egg Rock, Muscongus Bay

Islands as Biological Communities

David Quammen

PETER RALSTON

E
ver since Charles Darwin got back from the Galapagos, biologists have realized that islands produce more different species, and species that are more extravagantly different, than mainlands do. Islands are havens and breeding grounds for unique and anomalous species of all sorts. They are hothouses for wild evolutionary innovation. Why is that? The basic reason is isolation. Geographic isolation is a crucial part of what makes the evolution of new species possible. Islands simply provide more situations of geographic isolation than do mainlands. That is completely self-evident.

The Galapagos Islands are especially famous for their endemic species, but that is really only because these islands are where Charles Darwin happened to stop in the course of the voyage of the BEAGLE and saw this phenomenon in the flesh. If Darwin had stopped in the Hawaiian Islands instead, which he could just as well have done, he would have seen the same sort of thing — evolution made particularly dramatic and manifest — and we would now hear all about Darwin's honeycreepers and Darwin's fruit flies instead of Darwin's finches. But [what is true of the Galapagos Islands] is equally true of islands all over the world; they are full of evolutionary oddities....

Within recent centuries, and probably throughout all time, islands have been death traps for species. Within the past 400 years, 171 species and subspecies of bird have been recorded as going extinct; of these, 155 were island forms of birds. That's about 90 percent, despite the fact that only about 20 percent of the world's bird species and subspecies are endemic to islands. On a global scale, it means essentially that island birds face about a 50-times-greater likelihood of extinction than mainland birds.

The most famous of all extinct island species was from the Mascarenes, a little cluster of three volcanic islands in the Indian Ocean: *Raphus cucullatus*, as it is known to science; the dodo, as it is known to the rest of us. There are no stuffed dodo specimens in any of the world's museums; all we have are a pile of bones that have been assembled into a couple of complete skeletons and some other bone fragments. The dodo is practically the poster child of extinction. What I mean by that is that it is the symbol of all the species that humanity has managed to eradicate because those species weren't quite as resilient or as adaptable as they might have been. . .

By about 1680, the dodo was extinct. But it didn't just become extinct; it very quickly became famous for being extinct. It captured imaginations in the European world, because it was big, ugly, helpless, strange — and because it was the first case in which humans realized that we ourselves

had caused the extinction of a species. A biologist named Carl Jones, who works on the island of Mauritius, has told me that he thinks that was a watershed moment in the dawning of human consciousness about our relationship with the natural world: We saw that a species was extinct, gone forever, and we knew we were responsible for that extinction.

So what is it about the island situation that entails this special jeopardy of extinction? The basic answer is simple: islands, because of their limited area, support only small populations of any given plant or animal species, and small populations are more likely to be wiped out. What wipes them out? Various causes. It might be an exotic predator that has invaded the island. It might be a volcanic eruption or catastrophic drought or a huge fire. It might be humans hunting the species, destroying its habitat. Or, more likely, it might be a combination of those factors, a combination of direct persecutions and unlucky accidents that pushes a population below its threshold of genetic viability. There are too few left for it to be a healthy, sizable gene pool. That leads to inbreeding and inbreeding depression.

The point about all of this is that all populations of animals and plants naturally fluctuate in size, from time to time, in response to the good conditions and the bad conditions that they encounter. In a series of good years — mild winters, plenty of food — a population tends to fluctuate upward. Then there might be a series of bad years — predators and competitors are abundant,

Great auks once nested on Maine islands

JOHN JAMES AUDUBON

198

PETER RALSTON (2)

Atlantic puffin returning with herring to feed its chick

disease outbreak, drought — and the population fluctuates downward. All natural populations fluctuate in size, from time to time, in response to these factors. And small populations — meaning, in particular, island populations — are more likely to fluctuate to zero when conditions are bad, because zero is not far away. Rarity itself is the foyer of extinction. And island species tend to be rare.

Islands stand as warnings about the processes that affect any small, isolated patch of landscape. That is a crucial truth nowadays, as all the world's mainland ecosystems are getting chopped up into small, isolated patches. We all recognize that humans have spread across the continents, conquering wilderness, leveling forests, draining swamps, building roads, slicing the world into pieces by means of clear-cutting and fence-building and road-building and all the other things we do in the name of civilization and development.

Where wild lands do still exist, they have, in most cases, been left isolated, as relatively small fragments of the great mainland ecosystems of past ages. What isn't so widely recognized is that these fragments — surrounded by our highways and suburbs, our rice paddies and our villages, our farms and golf courses and malls and drive-in movies and Starbucks franchises — are becoming ecological islands in an ocean of human impacts. This is a crucial truth because, as islands, they are subject to that island syndrome I have mentioned, that special jeopardy of losing species

to extinction.

Island biogeography is the study of what species live on which islands. But the phrase has a second sense that is slightly less literal, although just as important. In this broader sense, island biogeography encompasses the study of any isolated patch of landscape. A patch of forest surrounded naturally by grassland is an island as far as forest-dwelling species are concerned. A patch of swamp surrounded by airport runways is an island, if you happen to be a frog. A lake is an island, if you happen to be a fresh-water fish.

1997

Steve Kress placing puffin decoys on Matinicus Rock

199

Keeping Track of the Wondrous Machine

Ben Neal

Midnight. With a high cirrus deepening and thickening into layers, it will soon be a night without stars. The wind is only perhaps 15 knots, just having shifted to the southeast, but an intermittent high keening sound in the rigging hints that something stronger is on the way. The crescent moon appears sporadically, wreathed in fleeing clouds, and then is extinguished from sight.

It is the next-to-last day on the last leg of the autumn National Marine Fisheries Service groundfish trawl survey. My watch is just turning out for our six hours on deck.

The Fisheries Service has the difficult and often contentious task of managing and conserving the United States' valuable marine fish stocks. Providing an effective management strategy means having an understanding of the underlying physical and biological processes that control the abundance of these species, and to this end the Service conducts various annual at-sea surveys. Last October I was invited to participate as a guest member of the scientific staff.

The purpose of this trip was to assess the regional abundance of the Gulf of Maine's groundfish stocks. Estimates made from data collected on this and other similar trips will have a direct influence on management practices as they are used in the development of the yearly catch quotas, the length of season and determining open areas for fishing.

This survey, which dates back to 1885, is the longest-running fish assessment of its kind, and is universally recognized as one of the most scientifically valuable time series in the world.

The surveys are done on vessels operated by the National Oceanographic and Atmospheric Administration (NOAA). I was aboard the venerable ALBATROSS IV, built for this work in 1962, and

Ben Neal with a 51 pound female cod

COURTESY OF THE AUTHOR

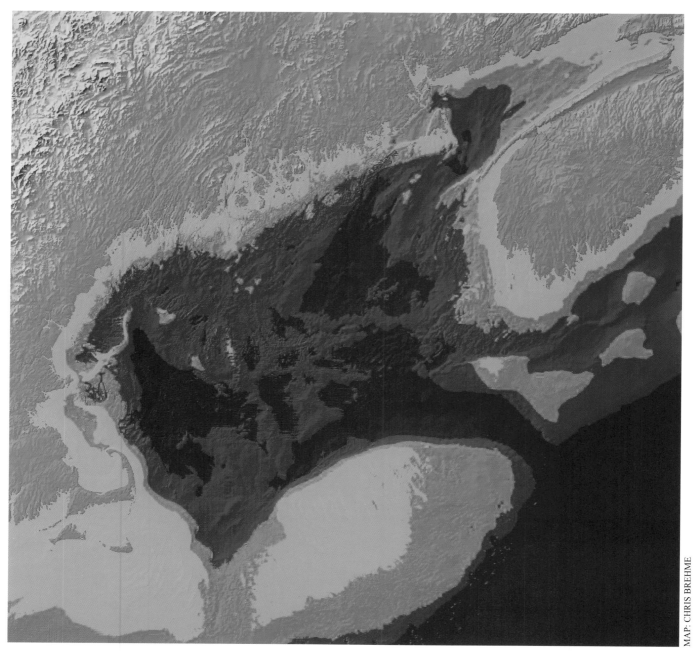

The Gulf of Maine by any definition can be described as a semi-enclosed body of water, with a complex bathymetry of banks, swells, ledges and deep-water basins.

MAP: CHRIS BREHME

which continues to this day to be the primary vessel for the trawl surveys. She is 187 feet in length and is equipped with stern trawl fishing gear, as well as plankton collecting equipment, processing tables, large freezers and a lab for preserving samples. The ship also provides accommodations and support for the 32 scientists, officers, crew and fishermen who work on board. Shipboard life is a routine of six hours on, six hours off. I work from midnight to just after dawn, have breakfast, then wake just before lunch and work until dinner. Obviously meals take on a special importance in this schedule, certainly as events to look forward to, but after a week they also serve to remind one of what time of the day it really is.

These standard groundfish trawl surveys are conducted twice annually, in the spring and the fall. Each survey consists of four legs, and each leg may have as many as 75 individual stations, each consisting of a 30-minute tow of the net at a prescribed speed (3.85 knots), with the entire catch coming on deck for sorting, identification and dissection by the scientific crew. The sex and maturity of fish are noted, and scales, otoliths (small bones in the heads of fishes, used for determining ages of individuals) and stomachs (for determining feeding ecology) are collected. Whole fish also may be collected and frozen for later shoreside use.

Sampling requirements are dictated by individual requests; on my cruise there were 33 separate requests, mostly from fisheries service scientists, but also from universities and research groups as far afield as Great Britain and Wisconsin. CTD (Conductivity, Temperature and Depth) measuring equipment is lowered at each station, and at selected stations a twin plankton net also is towed to assess primary productivity. In this way the ecosystem is measured from the lowest level of the food chain.

In my kitbag I carry an invaluable reference work, *Fishes of the Gulf of Maine*, by Henry Bigelow and William Schroeder, commonly referred to as "Bigelow and Schroeder." Published in 1953, this 577-page volume is still considered an authority on topics such as the range of species, their breeding habits, abundance and physical description. The book is noted for its accurate documentation of historical sightings (going as far back as John Smith's *Generall Historie of Virginia, New England and the Summer Isles*, 1616), its evocative and expressive line illustrations and vivid prose (at least, to those interested in fishes), and I keep my old copy handy on deck for any questions I might have.

The Gulf of Maine is described somewhat loosely by Bigelow and Schroeder as "the oceanic bight from Nantucket Shoals and Cape Cod on the west, to Cape Sable on the east, thus it includes the shorelines of northern Massachusetts, Maine, and parts of New Brunswick and of Nova Scotia." The National Marine Fisheries Service, for the purpose of its survey work, takes a more limited definition, and does not include Georges Bank or Canadian waters in the scope of our trip. (Georges Bank is covered by another leg dedicated especially to that rich and vast area, and the Canadians do the same for their waters.)

The Gulf of Maine by any definition can be described as a semi-enclosed body of water, with a complex bathymetry of banks, swells, ledges and deep-water basins. First noted on 16th century French charts as the "Sea of Norumbega," these waters contain some of the most prolific fishing grounds in the world, grounds that have stood nearly four centuries of intensive fishing for cod, haddock, hake, flounders, halibut, redfish and other finfish, as well as crustaceans and shellfish. Cape Cod, Browns Bank, Jeffreys Ledge, Stellwagen Bank and Cashes Ledge are all well known to generations of New England fishermen, and it is in this fruitful and storied region that we make our sampling tows.

We depart from Cape Cod, ranging east towards the Hague Line separating Canadian and American waters, then north to the Maine coast, and finally west and south to Cape Cod Bay, returning to Woods Hole through the Cape Cod Canal.

There are over 200 species of fish and shellfish native to these waters, of which only 40 or 50 are harvested commercially. We generally catch only the demersal, or bottom-dwelling, inhabitants. We do bring up some quantity of the vast shoals of herring and mackerel that serve as fodder fish for much of the rest of the ecosystem, but we do not encounter any of the larger, highly migratory predators roaming the Gulf. The giant bluefin, the world's fastest and most valuable fish, is a yearly summer visitor to the Gulf, arriving when the water temperature is above 50 degrees and schools of herring and mackerel make the living easy. Bigelow and Schroeder credit a 1,225-pounder brought into Boston in 1913 as the official record, but hint at rumors of 1,600-pounders. Bluefin of any size are rare in autumn; globetrotting adventurers that they are, these powerful fish potentially could be wandering among the West Indies, sliding through Gibraltar into the warmth of the Mediterranean or basking in the sunny Azores. Quarter-ton swordfish and ten-foot great white sharks also could be circling beneath us in the water column, but these "apex" fish are not our quarry.

Our interest is in the smaller fishes of the bottom, the numerous and productive species that make up most of the biomass, like the blades of grass that sustain the terrestrial world. It is a good thing that we are not looking for the giants, for they are more elusive now than they ever were, as their tribes regretfully have been reduced to only fractions of what they once were.

One of the more commonly seen fish on our trip is, of course, the famous and depleted Atlantic cod. We generally catch a basket or so of codfish on most hauls, for perhaps a total of 50 pounds, mainly scrod and medium-sized fish. There are a few individuals this night weighing up to about 15 pounds, but the average is probably well under five. We measure and dissect the fish, noting sex and reproductive state. We check their stomachs for their contents, remove their otoliths for further aging study back in the lab, and return the rest to the sea.

Cod were the fish that brought the first distant water fishermen and European settlers to the shores of the new world. Around the first millennium, Viking voyages, probably not coincidentally, covered much of the range of the Atlantic cod. Well before Columbus, Basque fishermen were marketing large quantities of salt cod in medieval markets, keeping silent on the location of their fishing grounds. After 1500, waves of English, Dutch, Portuguese and French immigrants to the New World began pulling boatloads of codfish from the sea. The easily preserved, low-fat cod continued to be the mainstay of the commercial fisheries in the Gulf of Maine until ice became plentiful in the mid-19th century, when fresh haddock was subsequently welcomed in the markets.

Sometime in the darkest hours of the night, with the bright sodium lights and the fishing routine focusing the consciousness down to little else than the patch of wet steel on which we would dump the next haul, from the net spills out the largest cod I have ever seen. Fifty-one pounds and almost four feet long, with stocky shoulders on a swelling olive-green back speckled with striking yellow spots, and a snowy white belly, this fish seems too large and heavy out of the water to flop about like its smaller brethren. She (large fish of this size are surely females, a fact we later confirm by examining this fish) lies quite still, with the placidity that seems to be common to the largest individuals of any species. The size of her head brings to mind the various codhead recipes I have seen in older cookbooks ("take one large Cod heade …"), and makes me understand how such a head truly could be expected to provide sustenance for a whole family. I reach down and lift her with both arms, close to her deep amber eyes, larger than my own. We put the mighty fish in a basket of its own, and put her on the scale.

After the haul I pull out my battered Bigelow and Schroeder and find that this fish is likely at least ten years old, and that while she is the queen of this particular trip, she is far from being a record for her kind. While most cod are caught when they are under ten or 12 pounds, the species can attain tremendous sizes. The largest on record dates from May 1895, when a longliner off the Massachusetts coast brought up a six-foot giant weighing 211-and-a-quarter pounds; another is noted as tipping the scales at 138 pounds dressed (it must have been over 180 pounds live) caught by a handliner on Georges Bank in 1838. Another of 100 pounds is recorded from Wood Island in Saco Bay, taken April 9, 1883, and noted as having a 17-and-a-half- inch head. The largest "recent" reference in the book dates from early July 1922, when a 90-pounder was taken in Maine coastal waters. However, then as now, any fish over 70 pounds is exceptional, and large fish are considered to be any over about 20 pounds. Our fish that night elicits such admiration even from the seasoned fishermen that she is not sent back to the sea, but rather split and cleaned for salting by one of the Portuguese fishermen in the time-honored fashion.

The common codfish is quite uncommon in its incredible fecundity. It has been said that if all the eggs of spawning cod survived to maturity, they would displace all the water from the Gulf of Maine within three years. Bigelow and Schroeder

note a 52-and-one-half-inch fish, roughly equivalent to the fish we took that night, which yielded 8,989,094 eggs. A more common three-foot fish could produce three million, and an average ten-pounder at least a million.

Unfortunately, this productivity was still not enough to keep the population of cod ahead of human fishing pressure. Decades of intense fishing by modern fishing trawlers have caused a drastic decline in groundfish abundance, a decline that may or may not have yet reached rock bottom. This decline has led some observers to question whether or not some stocks can ever recover. I take it as perhaps a good sign that most of the cod we see are juveniles, waiting to rebuild the population if they are able to breed before they are caught.

Not all of the fish that appear out of the net during these midnight hours are so photogenic and pleasant as the cod. One haul brings to our sorting table the writhing hagfish, by all appearances a worm of the bottom, grayish in color, scaleless, about a foot and a half long, without eyes or jaws or even true fins. This cartilaginous fish lives in the mud, scavenging whatever carrion drops from above. They are reviled by fishermen for eating their way into hooked or netted fishes, boring into the body cavities and eating out the intestines and the meat, leaving nothing but an empty sack of skin and bones. Their preference for the valuable haddock does nothing to improve their status. They complete their unpopularity by being prodigious producers of loathsome slime. Mucus sacs on either side of the abdomen can pour out this slime in quantities out of all proportion to the fish's small size, and Bigelow and Schroeder note that they do not consider reports of one hag filling a two-gallon bucket with slime to be an exaggeration. Luckily we catch few, and they seem well behaved, although not at all pleased to have been brought to a cold and windy deck in the middle of the night.

Another common sight on the sorting table are the toothy, gaping jaws of the monkfish. This creature is officially called the American Goosefish, but is also known as monkfish, angler, poor-man's lobster, molligut, all-mouth, the fishing frog. Out of the water his soft body collapses, and somewhat resembles the flattened skate. The huge head and mouth make confusion with any other species unlikely, however. Almost half his body is head, and his mouth, which he insists on holding wide open, stretches the full width of his body. The rest of the body tapers off to the tail, where the only edible portion of the fish is to be found. This tremendous mouth serves to support the monkfish's vast and storied appetite. We examine the stomachs of those we catch and find all manner of finfish, some almost as long as the specimens we

NOAA

We generally catch a basket of codfish on most hauls, for perhaps a total of 50 pounds, mainly scrod and medium-sized fish. There are a few individuals this night weighing up to about 15 pounds, but the average is probably well under five.

are examining. I read up later on the fish and am surprised to find a vast recorded diet, listing documented examples of just about every common species of fish having been found in the stomachs of this equal-opportunity predator, as well as accounts of finding snails, grebes, ducks, loons, herring gulls, lobsters, sand dollars, sea turtles, hermit crabs and even eelgrass in their stomachs. Bigelow and Schroeder cite one particularly full monkfish as containing "21 flounders and a dogfish, all of marketable size."

Late in the night we take a small alligatorfish, an odd little fish armored with bony plates. This rigid covering gives him the odd industrial form of having an octagonal trunk tapering midway to hexagonal. A source in Bigelow and Schroeder describes him as "not much thicker or softer than an iron nail." I find him as we clean out the sorting

box, his three-inch sticklike body hiding under a much larger pollock. Regardless of size, all creatures are measured, weighed and recorded, and thus I duly place him on the dissection table safely out of the wind that could blow him away.

My shift ends at six o'clock, and the second shift of scientists, whom I only see at these moments, arrive on deck. The wind has strengthened through the night to over 30 knots out of the southwest, and it finds damp and cold pathways to the skin at the neck and cuffs. Seas have built and salt spray mixes with a sharp blown rain; we slide buckets of fish across the wet deck, rather than trying to lift and walk with them. I welcome the thought of the warmth of my berth....

2000

Free Willy:
Eat or Be Eaten?

Colin Woodard

A cold wind whips across the cliff-bounded bay on the island of Heimaey, Iceland, which is now covered in whitecaps. But Keiko the whale could care less. Circling in his enormous floating pen, the 10,000-pound star of the 1993 film *Free Willy* ignores wind, waves and even the puffins and fulmers soaring overhead. His attention is focused on the bright orange boat that's just tied up alongside the pen. On every lap around the soccer-field–sized pen, he pops his head high out of the water to check out his latest visitors, looking very much like an excitable dog welcoming his human family home at the end of the day.

"Please, just try to ignore him," one of his trainers asks. "He recognizes everyone by sight and can get very excited when somebody new shows up."

At close quarters it's not an easy thing to ignore a 22-foot orca whale, particularly a trained one like Keiko. I'm promptly escorted to one of the floating pen's equipment shelters, which has an open door facing the gregarious whale. We remain in the shadows, where I am supposed to concentrate on interviewing the whale's trainers. It's hard to keep a straight face while standing there, dictaphone in hand, "ignoring" the whale. Every 30 seconds or so Keiko comes around for another pass just a few feet behind me, letting out thunderous exhalations as he lifts his head out of the water to peek inside the shelter.

But his guardians, the Santa Barbara–based environmental group, Ocean Futures, have asked me to give Keiko the cold shoulder. They're trying to fulfill a promise to "Free Willy," reintroducing the orca to the Icelandic waters where he was caught more than two decades ago. But Keiko was only two years old

GARY COMER

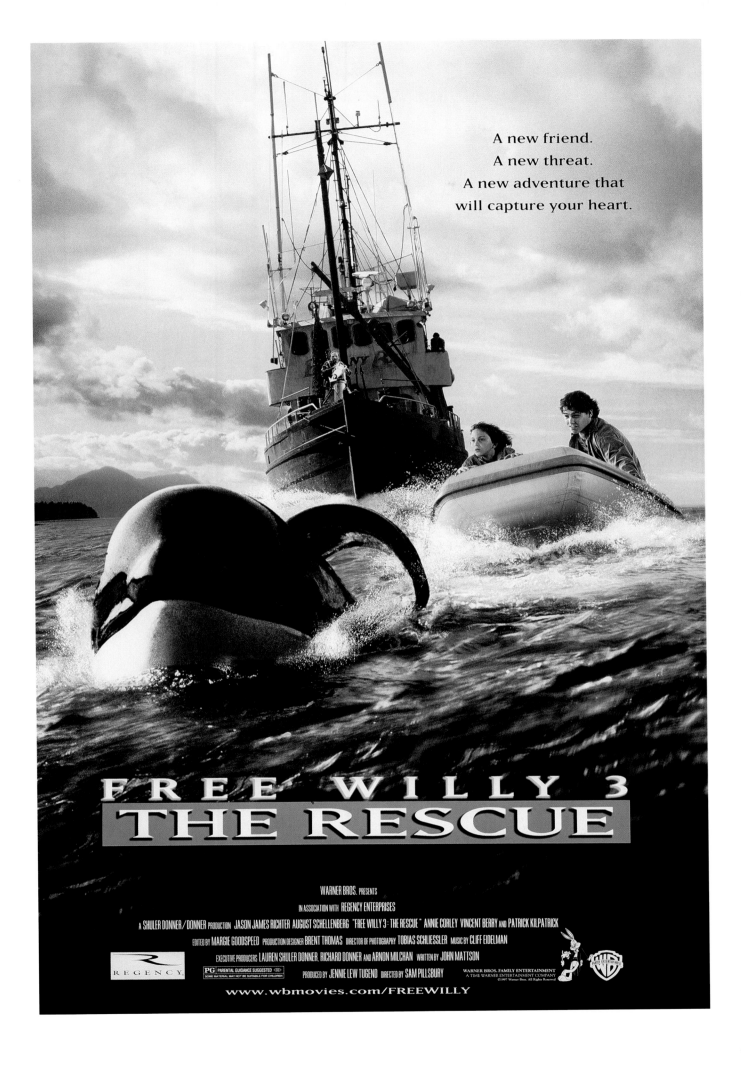

when he was taken from the wild. He's 23 today, and those intervening decades have been spent performing, entertaining and depending on humans. His trainers are trying to weaken his obvious bonds with people and encourage him to exercise, hunt and eat native fish, and explore the undersea environment.

But right now he's more interested in socializing with the media.

Occasionally I glance back and meet Keiko's curious stare and can't help feeling like I'm in the presence of a very big, very friendly, very intelligent family dog.

"You can't help but get a feeling of connection when you're with a whale," Charles Vinick, Ocean Futures' executive vice president, is saying. "They are thinking in some clear way. You get into the water with them and you say that there's a special connection going on that's almost physically tangible."

OCEAN FUTURES

"They're very special animals," adds research director Jeff Foster, who's spent much of his life working with captive whales. He says orcas combine feline aloofness with at least canine intelligence. I'm wondering how the family dog would fare if returned to the wild. All the dogs I've ever owned would be begging at somebody's doorstep within a day or two. Will an orca be any different?

During this March 2000 visit, the trainers are busy trying to interest Keiko in chasing the live cod they've been releasing into his pen. They have large plastic tubs full of live cod, hovering in some confusion in the middle of their restrictive plastic ponds, perhaps sensing the adjacent whale through some gaddian sixth sense. The cod were purchased daily from licensed fishermen here on Heimaey, the largest of Iceland's rugged and remote Westmann Islands. Before cod, Keiko was hand fed high-grade frozen fish. It's a considerable

expense. Keiko eats about 100 kilos (260 pounds) of fish every day. But food is just the tip of a dorsal fin when it comes to reintroducing a movie star whale to the wild. After starring in *Free Willy*, Keiko was discovered languishing in a poorly run Mexico City aquarium. The public outcry led a number of entrepreneurs to pledge to free the whale for real. Keiko was flown to Oregon where he spent months in rehabilitation. In September 1998 he was flown to Heimaey in a specially modified C-17, which itself suffered $1.5 million in damage to its landing gear on arrival. He now has a full-time staff of 12 who fly in and out of Iceland to tend to his needs in a massive fenced-in bay, where equipment is often damaged by storms. Ocean Futures has spent about $14 million on the project to date, and continues to lay out $1 million a year. They've gotten Keiko to take "ocean walks" behind their boat, but his encounters with pods of wild orcas have been standoffish. Project staff admit that it's entirely possible Keiko won't adapt to life in the wild, but they're going to give him at least another summer to make the proverbial jump to freedom.

The whole undertaking perplexes many Icelanders. Here humpback, sei, fin and minke whales have been hunted for centuries and are cherished not as individuals, but for their lean red steaks. While Americans have spent millions to rehabilitate a single whale, the Icelandic government has been threatening that it will leave the International Whaling Commission and resume commercial whaling. Polls show these policies have the support of more than three-quarters of Iceland's 270,000 people.

"Whales are a renewable resource that we want to manage and harvest in a sustainable manner," says Johann Sigurjonsson, a whale fishery scientist and director of the Marine Research Institute in Reykjavik. "We don't accept that some animals in the ecosystem are holier than others."

This is the fundamental philosophical divide between whaling and non-whaling nations. The arguments are often cloaked in conservation rhetoric, framed in terms of whether sustainable whale hunting is possible for a given place or species or using a particular technique. But the real difference, rarely spoken, is over whether a given species of whale should be hunted at all.

Many people in the United States and other countries believe the great whales simply shouldn't be hunted at all. Whales — by dint of their magnificent size, appreciable intelligence and noticeable sociability — are regarded as fundamentally different from a codfish. Keiko, the most famous nonfictional whale in human history, is the subject of considerable public interest, with tens of thousands following his progress on the Internet. Every

Before a worldwide moratorium went into effect 11 years ago, 90-ton fin whales were harpooned, lashed to the side of the ship and hauled individually or in pairs back to Iceland's single whaling plant in Hafnarfjordur.

piece of Ocean Futures merchandise, every press release and business card, features two names, side-by-side, beneath the logo: organization president Jean-Michel Cousteau and Keiko. Keiko, like many whales, has been inducted into humanity.

Most Icelanders don't see it that way. Whales are an exploitable natural resource like cod or herring, and their stocks should be carefully managed and sustainably harvested. Iceland even has an impressive record when it comes to fisheries management. Iceland fought several near-wars with Britain to protect its fisheries by creating the world's first 200-mile Exclusive Economic Zone. Subsequently Canada and the U.S. fished their cod stocks into near-oblivion, while Iceland's cod fishing industry remains relatively healthy due to careful scientific monitoring and strict quotas.

Sigurjonsson points out that based on the latest surveys, populations of fin, minke and sei whales are sufficiently healthy to sustain an annual catch of 100 to 200 of each. "Today I don't think [the anti-whaling argument] has anything to do with environmental conservation," he says. Based on the figures, Iceland's parliament has voted to resume whaling, but the government is moving slowly on the issue for fear of a boycott by major trading partners….

Nobody hunts orcas for food, so Keiko's prospects in the wild will largely be determined by Keiko and his trainers….

When we left the floating pen, the wind was starting to stiffen. Waves exploded into spray on the 27-year-old lava flows that form the other side of the island's harbor. Twin volcanic cones loomed above the sea — one created centuries ago and covered in grass, the other, a heap of steaming red gravel, was created in 1973. The ocean beyond looked harsh and forbidding, a place where ice and lava flows sometimes meet in crunching, hissing collisions. I wondered what Keiko makes of it — if he thinks about such matters at all — whether he'd prefer life out there to his carefree captivity in a U.S. aquarium.

As we powered our way across the harbor, I asked Jeff Foster what would happen if Keiko decides he prefers captivity to the chilly waters of southern Iceland. "We're not taking any chances with Keiko's life," he said. "If need be, we're prepared to take care of him for the rest of his natural life."

2001

CHAPTER EIGHT

Islands Far and Gone

THE *ISLAND JOURNAL* DEFINITION OF "FAR" has expanded over the years, beginning with the recognition that the islands of Maritime Canada, particularly Prince Edward Island and Grand Manan, had much in common with the island communities of Maine. By the mid-1990s the editors were looking for parallels further afield — in Ireland, where a resettlement effort in remote areas showed promise; in Scotland, where residents of Eigg in the Hebrides bought out their landlord and took control of their own destiny; on Baffin Island, where three young Mainers explored the newly established Native territory of Nunavut. At the invitation of the owner of TURMOIL, a private exploration vessel that ranges the globe, Institute president Philip Conkling visited and wrote about the Pacific coast of the former Soviet Union, the Aleutians and the Arctic Ocean. In 2002 writer Katie Vaux visited Sable Island off eastern Nova Scotia, home to several hundred wild horses and a woman who has studied them for three decades. The geographic reach of *Island Journal* is truly global because the concerns of islanders, no matter where they live, are so often the same.

David D. Platt

Opposite: Gary Comer *North coast of Spitzbergen*
Above: Peter Ralston

A Black Hole on the Map of Hope

Cabot Martin

I've tried at different meetings to express what the fishing moratorium in Newfoundland is like, but I guess the best way to describe what it's like is the fact that in Newfoundland today, even after a year, many people cannot talk about it. They find it hard to articulate their feelings. People who have spent their whole life talking about nothing else but fish don't want to talk about it. They'll talk about hockey scores, crises in Bosnia — about anything but the loss they are going through.

So what is the loss? First, it is the recognition that, given the power structure, we fishermen were — and are — helpless. As a group, we are political eunuchs. We have no power.

More fundamentally, we know that we have allowed great damage to be done to the one thing that kept our society together — the ecosystem that we were living off. If you ask older fishermen in Newfoundland, they will tell you that they don't expect to see a commercial fishery in Newfoundland ever again, that it is gone forever. There are others — and I am included among them — who take a more optimistic view of the tremendous healing powers of the ocean. But whether we will be given a second chance is a question over which we have no control. It's a process that must surely teach us something; if not, then we are incapable of learning.

One lesson, surely, is the need to build effective bridges between the scientific community and the fishermen, so that the two kinds of knowledge work together and expand their reach. It is critical that

JOHN DE VISSER

CANDACE COCHRAN

this be done in such a way that both groups respect one another in their different knowledge. We also need effective organizations that bring fishermen and processors together. I framed it that way because in Newfoundland there has recently been such a terrible class struggle between fishermen and processors that it's very hard sometimes to get the two groups even to sit in the same room.

This sense of working together has to expand to include the larger community as well. We have to recognize that people working on the water — fishermen — are a minority within this society. They will never be effective in exercising a wise stewardship of the ocean if the larger community doesn't value the ecosystem and work with them and respect them. So whether you like it or not, the general community and the fishing community

are going to have to work together.

The other thing we realize, in retrospect, is how naive and ineffective we were at lobbying government. The decisions were being made not in St. John's but in Ottawa, where the two trawler companies had full-time vice presidents of public affairs who were buying lunches for every influential bureaucrat they could get their hands on while they were holding press conferences locally.

But the fundamental lesson we've learned is that you can never take just the short-term view. In light of our modern human ability to destroy ecosystems, you always have to accept short-term pain for long-term gain. In Newfoundland we never were ready to bite the bullet. We always put it off. We learned the hard way. I suspect, if we look at environmental degradation in any other cat-

CANDACE COCHRANE

egory, that's a given. It's not some faraway problem in the Amazon jungle; we are living through it in a so-called modern society like Canada's. Yet we still are not prepared to take the short-term pain.

You hear so much in the media — and it's something of an assumption in our North American society — that environmental protection and economic growth are somehow opposites that clash; that given the level of unemployment, you have to come down on one side or the other. But that is a false dichotomy. There is no such conflict. We in Newfoundland are living proof that, because we did not put the environment first, our economy suffered terribly. We may have fatally injured our whole society because we didn't respect the environment.

The lesson of the cod in Newfoundland is just one example of the general lesson of our failures of stewardship all over this whole planet. We just happen to be an obvious example, where cause-and-effect is so clearly visible. But I suspect I could go to any country in the world and find the same short-term trade-offs, the same lack of stewardship. So if you can draw some parallel or some lesson from our experience, then perhaps some good can come out of it.

As for ourselves, I think we will survive as a society and even be a better society because of it. Right now, a lot of work lies ahead of us. I personally look forward to strengthening links between New England and Newfoundland, drawing on your experiences, as you perhaps can draw on ours.

1994

Lost Generation

Nancy Griffin

Newfoundland's settlements cling to the shore around the island's many bays the way periwinkles cling to a rock — tenaciously and often at angles that appear haphazard, almost dangerous. But there was nothing haphazard about the siting of these "outports," as Newfoundland coastal villages are called. They were designed hundreds of years ago for cod fishing — designed for the men who fished dangerous seas in small, open boats and the people on shore who split, dried and salted the fish.

Houses were simple, small enough to heat with a minimum of firewood, hunkered down against the ever-present wind, set close enough to the sea for access and far enough back to avoid its ravages.

The fish flakes and stages have largely disappeared now, since refrigeration allowed the building of freezer plants and changed the fishery forever. Most outports survived and adapted to this and other changes in fishing, processing and economics through the years.

While outport people have survived hard times, hard weather and the sorrow of losing men to the sea, they may not survive the biggest blow ever — the cod moratorium. Since the shutdown of the island's major economic resource more than five years ago, many thousands of reluctant Newfoundlanders have fled the outports for the mainland to find work. The exodus threatens the very existence of many small, closely knit communities since their future, the young people, are the first to go. Optimists wait, however nervously, expecting or at least hoping the cod will return and the fishery and plants will reopen and save their communities. Skeptics look back to the 1960s, when a government-mandated program forced year-round residents off most of Newfoundland's islands. They read into present government policies a subtle, unstated sequel to the hated resettlement process, which broke the hearts of people forced to leave communities settled by their forebears generations before. Pessimists (some would say realists) clench their jaws and pack their bags. "We got hit hard," said Leo Bruce of Placentia. "It's hard when the young people leave."

Bruce, a retiree, is a member of the board that runs the town's Star of the Sea Hall, a Catholic nonprofit club where most town events are held. During a regular Saturday night soiree in December, more than four dozen people of all ages danced to the music of Newfoundland and Ireland played by a local band.

"We used to get 300 people at this dance," said Bruce, looking around the darkened hall's large, uncrowded dance floor. "We'd have to lock the doors at 10 o'clock to stop letting people in. Another club in town has shut down."

This night, two of the dancers are "home" from Windsor Falls, Ontario, to celebrate their 25th wedding anniversary. No matter where they roam or how long they live away, Newfoundlanders always call The Rock home.

Another dancer, Cyril Stewart, moved to Placentia from Rushoon only a year before, but planned to move again in January, this time to Ontario to look for a job on the oil rigs. "I'd sooner be here. It's my home," said Stewart, a displaced fisherman. "But I'm not alone. Everyone's going to have to go."

When a Newfoundlander says "fish," he means cod. Atlantic cod, *Gadus morhua*, caused the island to be settled and sustained its residents for centuries. On July 2, 1992, the 500-year-old fish-

CANDACE COCHRANE

ery closed abruptly, in mid-season, when fisheries biologists announced they had overestimated stocks by more than 100 percent. Since then, unemployed fisheries workers have been supported by various programs colloquially dubbed "the package."

For centuries, the rocky, cold and windswept island's bountiful codfish supply provided subsistence for most residents and riches for a few. For decades before the moratorium, cod fishing and processing offered employment at decent wages for many. The reopening of a commercial cod fishery is not in sight and outports are emptying as workers head for the mainland, seeking stable, secure employment — something they fear they may never see again in Newfoundland.....

Unemployment and outmigration are hardly new phenomena here, but Newfoundlanders are terrified by the scope of today's exodus and extraordinary lack of opportunities. Experts predict the population will decline by 60,000 to 500,000 within the next few years. But it's hard to track all the one-way U-Haul rentals and guess which ferry passengers won't buy a return ticket, so some observers believe the population has

already dipped near or below that figure.

The signs are there. Songs about empty nets and survival of outports dominate the music scene. Provincial television shows air increasingly frequent stories about Newfoundlanders leaving home. "There's no question the face of rural Newfoundland will be altered, hardly recognizable in 10 years," said Jim Wellman, who retired last year as host of the nightly "Fisheries Broadcast." "The Broadcast," as islanders know the 47-year-old Canadian Broadcasting Company radio show, indicates the importance of fish to Newfoundland. It's the longest-running live daily radio show in North America.

The former host predicts the fishery will never be the same, even if the cod return. He believes bigger boats, not traps and open boats, will dominate the new cod fishery. and this means fewer fishermen, fewer plants, fewer communities.

"I hate to see it," said Wellman. "The essence of what is Newfoundland is going to die, because economically it makes sense. I believe the politicians are finally beginning to realize what they have done."

Newfoundlanders are deeply involved with

JAMIE LEWIS

CANDACE COCHRANE

their culture. Outports represent the most overt symbol for this unique heritage, which includes an approach to life still dominated by a healthy dose of humor. Wellman participated in a "Kitchen Table" panel discussion convened by educators last fall in St. John's to discuss the current crisis. The title: "Whither Newfoundland Culture? The arse is out of 'er. She's gone by's, she's gone." Wellman said he would explain the title to mainlanders who don't speak the island lingo as, "The ethnographic exploration of the interplay between piscivorous and archipelagic factors in Newfoundland: a comprehensive inquiry."

As for the moratorium, many fishermen believe it was not a moratorium on cod fishing at all. They believe it was a moratorium on Newfoundlanders fishing for cod. Beyond the 200-mile limit, foreign vessels still fish for anything they can find. Inside the 200-mile limit, freezer trawlers from Canada and other countries fish for other species under agreements with Ottawa. Ten percent of their catch may be cod "bycatch." One trip's bycatch would sustain a small-boat, inshore fisherman for a year.

In Newfoundland, "rural" means the same as "coastal," since few people live in the province's interior. Nearly half the island's residents live in and around the capital, St. John's. The other "livyers" reside in 1,600 tiny outports that dot the convoluted coastline. An outport, by definition, is a community settled by boat. Most outports remained accessible only by water for many years. In a speech at the Summit of the Sea conference in St. John's last fall, Dr. Harris described John Cabot's first glimpse of The Rock, probably at Bonavista, when the explorer claimed New Founde Lande for England:

"We can at least be certain that the coast he first encountered was a rugged one of beetling cliffs and torturously deformed ancient rock, pushed and crumpled and folded by continental collision, ancient vulcanism, and like titanic processes."

Combined with those factors, "the action of the sea has created an incredibly indented coastline with coves and bays and bights and sounds and tickles and fiords in number almost beyond counting," said Harris.

It was in the tiny folds and pockets of these bays and tickles that outports grew, sustained and supported by the cod in their waters.

"The outport cannot survive without fish. Either there are fish, or the community dies. There is no middle ground," said Harris. "The handwriting is on the wall. The outport as it was has all but disappeared and will not be recalled. To deny the reality of change is as foolish as it is pointless...."

1996

I, MILTON ACHORN

I, Milton Achorn, not at first aware
That was my name and what I knew was life,
Come from an Island to which I've often
returned
Looking for peace, and usually found strife.

Till I came to see it was no pocket
In a saint's pants while outside trouble reigned;
And after all my favorite mode
Of weather's been a hurricane.

The spattered color of the time has marked me
So I'm a man of many appearances;
Have come many times to poetry
And come back to define what was meant.

Often I've been coupled, and often alone
No matter how I try I can't choose
Which it shall be. I've been
Ill-treated, but often marvelously well-used.

What's a man if not put to good use?
Nothing's happened I want to forget.
What's a day without a notable
Even between sunrise and sunset?
My present lover finds me gentle
So gentle I'll be in my boisterous way.
Another one was heard to call me noble.
That didn't stop her from going away.
To be born on an island's to be sure
You are native with a habitat.
Growing up on one's good training
For living in a country, on a planet.

Shall I tell you the soil's red
As a flag? Sand a pink flesh gleam
You could use to tone a precious stone?
All its colors are the colors of dreams.

Perhaps only the colors I dream
For I grew under that prismatic sky
Like a banner of many colors
Alternately splashed and washed clean.

PETER RALSTON

The Island's small … Every opinion counts.
I'm accustomed to fighting for them.
Lord I thank Thee for the enemies
Who even in childhood tempered me.

I beg pardon, God, for the insult
Saying You lived and were responsible
… a tortuous all-odds-counting manner
Of thinking marks me an Islander.

Evil's been primary, good secondary
In the days I've been boy, youth and man.
I don't look to any rule of pure virtue
But certainly not continuance of this damned
…
Damned! Damned did I say? This glorious age
When the ancient rule of classes is hit
And hit again. History's greatest change
Is happening … And I'm part of it.

Milton Achorn, *1991*

221

"Up For Sale"

Maggie Fyffe

COURTESY OF THE ISLE OF EIGG HERITAGE TRUST (5)

The Isle of Eigg has a colorful and often turbulent past, but what has happened in recent years will undoubtedly be recorded as a major landmark in recent history. Before Keith Schellenberg bought the island in 1975 (snatching it from public ownership by narrowly outbidding the Highlands and Islands Development Board), the population had dropped to a dangerously low level, so when several young families (ourselves included) were brought in to do a variety of jobs, we were welcomed by the indigenous islanders. I was delighted to live in such a beautiful place, to have a safe environment to bring up children and to be part of a small community — the politics of landlordism had yet to make an impression upon me….

By 1981, we were lucky enough to have bought a derelict cottage with attached croft tenancy and although it was a relief to have a secure place to live, it ironically coincided with my husband and several other local men being made redundant. There were constant injustices but in general people were unwilling to speak out in case it would jeopardize their house or job. Most people working for the estate

One thing we mustn't lose sight of is the fact that although we were successful, the land laws in Scotland haven't changed and many other communities are still suffering at the hands of unscrupulous owners.

felt they were being manipulated, but due to a chronic lack of independent housing and alternative employment faced the option of leaving Eigg or moving into a caravan [mobile home], which is what a substantial number of families did....

There had been frequent discussions about the possibility of community ownership but there certainly wasn't the consensus or the confidence to put these ideas into practice until 1991, when a group of people concerned about land ownership in Scotland formed the Isle of Eigg Trust and brought their ideas to the island for our consideration. Many meetings later, the Eigg Residents Association — which consists of every adult member of the community — voted in favour of supporting their aims. It's interesting to note that at this stage we were still looking to others to take the lead.

In May 1992, Eigg was officially put on the open market. Interest had been shown by the Scottish Wildlife Trust (who had had a presence on the island for a number of years) and the Highland Council (our local authority) in trying to form a consortium with the community, but time and funding were in short supply and we were unable to prevent Mr. Schellenberg from buying his ex-wife's share of the island....

Nineteen hundred and ninety-four was the year in which things changed radically. We woke up one morning in January to learn that Schellenberg's vintage Rolls-Royce had been destroyed by a mysterious fire. The resulting press coverage with Schellenberg's numerous libelous statements and his insistence on differentiating between islanders and incomers angered the community considerably and led to the indigenous population issuing the following statement: "We who have been born and brought up on the island would like to refute utterly the ludicrous allegations about the community here. The island has a small but united population of local families and incomers who are between them struggling to develop a community with a long-term future against the apparent wishes of an owner who seems to want us to live in primitive conditions to satisfy his nostalgia for the 1920s. If the nature of the island has changed, it could be something to do with the fact that all the local men working for the estate during Schellenberg's first years have left, taking their indigenous way of life with them. The incoming islanders play an active caring part in the community. They help run the senior citizens lunch club, they drive the community minibus to enable those without transport to get to the shop or church, and they have organised a Gaelic playgroup so that their children will have a chance of learning Gaelic in order to preserve the traditional culture of the island."

This marked an important change in our relations with the media — we made sure from then on that our views were represented….

Over the course of the next few months, a variety of rumors were in constant circulation — Eigg was up for sale, was about to be sold, had been sold! Stubbornly I refused to believe the stories, convinced that if something so significant to the people of Eigg had happened, we would surely have been told. Not so. In March 1995, we learned from the press that we had indeed "been sold" for £1.6 million to a mysterious German artist who called himself "Maruma." We issued a statement to say that we were relieved that the uncertainty and speculation of recent weeks was over, that we had felt restricted and suppressed by the previous management and that we looked forward to meeting Maruma, who we hoped shared our aims and ambitions.

Maruma arrived by helicopter one Sunday morning in April. He visited every household to introduce himself, and although rather stunned, we hastily arranged a residents' meeting for the following day. We presented him with a copy of our workshop results: Eigg as we saw it. We also raised the most pressing issues, i.e., lack of leases on both private and business premises and associated land and the urgent need to provide a site for a new community hall. He seemed to appreciate our concerns, promised to look into the matters we'd raised, talked vaguely about a concept for the island that would be formulated in consultation with the community and expressed a desire to be a part of the community and not to be seen as "owner."…

Alarm bells started to ring in July when we had an unexpected visit from lands agents from Knight Frank, who informed us that they were looking at the extent of the dry rot in the main house…. A week later, after much interest from the media, Knight Frank was forced to announce that Eigg was indeed "Up for Sale" — with a price tag of £2 million!

Because most of the groundwork had already been done, we were able to move swiftly. On August 17, with journalists from every major newspaper in Britain in attendance, we launched our public appeal. That's when the hard work really started. We set up a database with as many names and addresses as we could think of and started to send out leaflets. Likewise the Highland

The island has a small but united population of local families and incomers who are struggling to develop a community.

Council distributed the leaflets through libraries and other public buildings, and the Scottish Wildlife Trust produced a separate leaflet (with the emphasis more on conservation), which was sent out to members of all of the 47 wildlife trusts nationwide.

The response was totally overwhelming — letters of support and donations began to flood in from all over the world and the phones rang constantly. Not only was there total commitment from the people of Eigg but also from politicians who brought the matter up in Parliament, friends who organized fund-raising events, and musicians who organized benefit *ceilidhs* all over Scotland and as far away as Detroit!

We had set ourselves the target of raising £800,000 through the public appeal; the rest of the funding was expected to come from the National Memorial Heritage Fund. It was a huge blow when they announced that they would like to support us in principle, but were unable to do so unless our constitution was altered in favour of a majority for an appropriate conservation body, e.g., The National Trust.

Much debate followed. Were they really prevented from supporting our bid or had there been behind-the-scenes political pressure from a gov-ernment opposed to land reform?

The Isle of Eigg Heritage Trust had by this time become a legally established company with charitable status comprising four elected representatives from the community, two from the Highland Council and two from the Scottish Wildlife Trust with an independent chairman. We felt unable to alter this structure mainly because (a) all partners were committed to its present form and (b) it was on this basis that the general public had responded so generously to our appeal. As the closing date for bids approached, salvation came in the form of an anonymous donor who wished to pledge £750,000!

On November 28, Mark Cherry, chairman of our residents' association, was followed by television cameras as he made his way to the office of Knight Frank in Edinburgh to hand in our bid of £1.2 million (well below the asking price but based on a professional valuation). Two days later we learned that our bid had not been accepted as it failed to reach the highly inflated asking price of £2 million. Although obviously disappointed, we were by no means dispirited; the fund-raising continued, helped enormously by the continuing interest and support of the world's media....

The moment we had worked towards for so

The official hand-over ceremony (left and above) included the requisite piper and the unveiling of a commemorative plaque.

long finally came at 4 p.m. on Friday, April 4, with a phone call to say our bid had been accepted! Within minutes everyone on Eigg had heard the news; a few minutes more, so had the rest of the world. It was at least four hours before I managed to get off the phone and drink a toast to the future!

The minute the celebrations were over it was back to work. We had until the official hand-over date of June 12 to write to everyone who had made a pledge to send the money as soon as possible. We needn't have worried — the pledges flooded in and much more besides, and the deal was concluded.

The official hand-over ceremony took place on June 12 and was certainly the biggest celebration Eigg had ever witnessed. Approximately 400 people from all over Britain braved the wind and rain to be with us as a commemorative plaque was unveiled by our two oldest residents. Politicians, councilors, conservationists and members of the community made emotional speeches, the primary school children sang a Gaelic song and some of Scotland's best musicians entertained us well into the early hours. A day to remember.

The Isle of Eigg Heritage Trust has since established a mechanism for management with

eight directors and the chairman meeting on a regular basis as well as spending long hours communicating by fax and phone. Most of the work done to date has been research on how best to proceed — every decision must be well thought out if it is to be of lasting benefit....

The rebuilding of Eigg's infrastructure will be a long, slow process but with a feeling of security and stability beginning to return, we are all looking forward to the challenges ahead. One thing we mustn't lose sight of is the fact that although we were successful, the land laws in Scotland haven't changed and many other communities are still suffering at the hands of unscrupulous owners. With a new government and a future Scottish Parliament, we must make sure that the subject of land reform stays on the agenda until suitable legislation is introduced.

1998

COURTESY OF THE JACKSON FAMILY (4)

Nunavut
A Journey Across Baffin Island

Hope, Dylan, and Tristan Jackson

Dear Outside Adventure*:*

I am writing in response to your Outside Adventure Grant. This is the adventure that I have in mind ... On April 1, 1999, Canada will return about 770,000 square miles of land, and the right to govern themselves, to the Inuit people. These people experienced great change in their lifestyle and culture due to the Anglo influence. Now they will have the opportunity to determine how they want to live in their land, to be called Nunavut.

This land is a hard land. I would like to visit these people and come to know a part of their heritage. To do this, I propose to fly into Pangnirtung on Baffin Island... From there we will explore the village, the nearby whaling stations and remnants of an ancient past, before beginning our walk across Auyuittuq National Park Reserve. A two-week hike across the park to Broughton Island will cover some 60 miles of Arctic terrain, giving enough time for side trips....

Hope Jackson at home on Green's Island, Vinalhaven

We live on a remote island off the coast of Maine. We have spent the last eight years living in a one-room dwelling with 12-volt/solar electricity, hauling water from a dug well and harvesting our own firewood for heating. For the past five years we have been homeschooled, and also have been the only year-round family on the island. We have been raised to be self-sufficient and self-motivated....

— Louisa Hope Jackson

This expedition we have planned is unique in that it is in an area very few people have ever heard of, let alone seen. The land is largely unaltered by human presence; last year only 1,000 people visited the park. I have always enjoyed the outdoors, in particular the wilderness, and I am attracted to the idea of going to such an isolated, unspoiled place.

— Dylan Jackson

...We plan to study by being keenly observant and through meticulous record-keeping. The goal is the education of our group and the sharing of this information on our return.

— Tristan Jackson

PREPARATIONS

Hope:

April 2 to 21, 1998: ...We have been gathering information for the last two weeks. The process up to this point has been very educational; we've had to do a lot of writing, making telephone calls and researching. We have been enjoying this part of the adventure, but we are definitely looking forward to the actual hike. It's going to be amazing!!

Dylan:

... I wasn't sure how I felt about our home being broadcast to midcoast Maine on the six o'clock news. In the morning, before the camera crew came out, Tristan, Hope and I had a quick talk to try and figure out what it was we were going to say to them. The meeting dissolved into us warning each other not to say anything too stupid. I think we did okay....

TRAVELING NORTH

Dylan:

July 22: ...The best aspect of the flight was the captain's translations. While describing the safety features the fact that English was his second language became apparent. After telling us what to do in French, and Inuktituk, he let the Americans know that we should tighten the seatbelts around our tiny lips — oops, that was supposed to be hips. If we had small children with us, he felt sorry for us. But we should put on our oxygen mask first, then help them; the same if we had those who acted like small children with us!

...We found out yesterday that a bowhead whale was killed on an island near Pangnirtung. And today they were going to bring all good eating parts to the landing at about 7:00 p.m. We were in our tents resting and heard lots of people cheering and talking. So we went over to the landing and practically the whole town was there to see the hunters come home....

Hope:

July 25: There were tons of people at the feast, and tons of whale pieces. There were a lot of speeches, but I couldn't understand anything they said....

Dylan:

...As tourism becomes a larger part of Nunavut's economy, it can be expected that outsiders' opinions will have more of an influence on how the Inuit live their lives. A hunter that I spoke with said that visitors are often disgusted at the sight of a seal being butchered on the beach. "I tell them to think of the last time they ate a hamburger. I tell them it's not so different; at least this seal was free before he died." I see the hunter's point, but many other people don't.

Hope:

Reflections: Four days ago, Pangnirtung was celebrating its first bowhead whale kill

"We crossed the Arctic Circle today... I've never seen a place like this before."

"Tommy would whip out his map and say, 'I think I've got it. Yep, we're on Baffin Island!' But Dylan, Tristan and I disagreed. We thought it looked a lot more like Mars than Earth. And it did…"

since 1946. There were over 230 strips of baleen in its jaw. The elders of this hamlet were given a pair each. A few days ago, we feasted alongside the rest of Pangnirtung's residents on a 40-foot-long table covered with whale blubber. Older Inuit women dressed in traditional clothing cut the skin off the blubber. The black skin, high in vitamin C, is the prime cut....

AUYUITTUQ NATIONAL PARK
Pangnirtung Fjord, Overlord Camp
Tristan:

July 28: Last night Jamie [our 17-year-old Inuit companion] decided that he didn't want to attempt the hike. I don't think that this decision was an easy one for him. From the time he arrived we could tell that, as much as he wanted to hike through the park, he also had a strong desire to spend the summer with his family on Broughton. This indecision worried the rest of the team because if there is one thing that an expedition requires, it is total commitment.

I think that Jamie would have made an excellent member of the team and I am sorry to see him go....

Hope:

Today at breakfast in our Overlord camp, Julia began to cry. She is so quiet that it was difficult to hear what she was saying. But her tears spoke clearly. She wanted to return to her home on Broughton Island. What a huge disappointment....

Reflections: There will be plenty of time for the team to sort out the lessons of our attempts with these two Inuit teens. Right now, they are focusing on the expedition and our remaining Inuit companion, Tommy. But questions about Jamie and Julia are running through my mind. Did we make them feel welcome? Were we clear in describing the commitment it takes to hike across Auyuittuq Park? Could the Jacksons have spent more time in the initial interviews? Regardless, we are now a group of six.

July 30:
Auyuittuq National Park: It is very windy

today. The worst is when the trail is sandy. Today we got sandblasted the whole way. It really hurts to be hit in the face by sand flying at 40 miles per hour. We crossed the Arctic Circle today... I've never seen a place like this before.

August 6: We have made it to our halfway point, Summit Lake campsite. It was wet, foggy and dreary all morning.... I have lost total track of time up here.... It's hard enough remembering what month and day it is, let alone the time.

August 9: Today we hiked up and down, up and down, over big old glacial moraines for five and a half miles. A few times we would stop and say, "Tommy, where in the world are we?!?" Tommy would whip out his map and say, "I think I've got it. Yep, we're on Baffin Island!" But Dylan, Tristan and I disagreed. We thought it looked a lot more like Mars than Earth. And it did...

Dylan:

August 18: We have been in the park exactly three weeks and we are all eager to get the boat to Broughton. The boat was hours late, but when it arrived, it was obvious why. There was a seal draped over the stern and another in plastic bags on the deck. The ride was only an hour and a half long.

1999

Sea of
Okhotsk

Kamchatka
Peninsula

Glimpses of
Lost Worlds

Philip Conkling

Along the margins of the chains of islands that circle the seas of the globe, marine life is always abundant. Whether we think of the fabled Galapagos with their thousands of diving seabirds and lounging marine iguanas, or humpbacks nursing calves near the shores of the Caribbean and Hawaiian islands, the images of abundant sea life are nearly synonymous with islands. This circumstance results partly from nutrients washing down the slopes of islands, fertilizing underwater gardens and nurseries where the critical life stages of marine life emerge, and partly from upwelling currents and swirling tides that keep these "littoral" or nearshore waters stirred into a rich and endlessly renewed broth....

When I first visited Alaska's Aleutian Islands in 1997, their wildlife seemed so rich compared to what

sia

Chukchi
Sea

Alaska

Nome •

Anchorage •

Bering Sea

Aleutian Islands

MAP: CHRIS BREHME

I had known in the Gulf of Maine. Two years later — in comparison with the stupendous abundance of seals, sea lions and seabirds we observed in the Kurils Islands between Japan and Russia— the Aleutians and their wildlife seemed suddenly paltry.

In the Aleutians each day, we saw hundreds to thousands of seabirds, a few score sea lions and three sea otters. In the Kurils, we saw hundreds of thousands of seabirds virtually every day, hundreds to thousands of sea lions and fur seals in many colonies and hundreds of sea otters. The differences were an order of magnitude more dramatic. Yet these two Pacific archipelagoes are both washed by cold Pacific Ocean waters of similar temperature. Both are surrounded by nutrient-rich upwelling currents, stirred by similar tides. The only logical explanation for the differences between them is that the Aleutians have been intensely exploited for their fisheries, especially during the past two decades, while the Kurils, by historical accident, have excluded virtually all fishing for most of the past century.

What lessons are we in the Gulf of Maine to draw from the Kuril and Aleutian island examples? The answers are not obvious. Here in the Gulf of Maine, where a rich fishing tradition has been an inseparable part of our history and identity for 400 years, no responsible person would seriously suggest we simply exclude all fishing (as in the Kurils) to re-create pre-colonial abundance….

In the Gulf of Maine, the marine stewardship task is too complex for the sort of two-sided confrontation that has become so typical of modern natural resource debates. There is no option of "restoring" a pre-colonial "natural ecosystem" here. That condition is irretrievably gone; we are simply missing too many of the major species, including virtually all the biomass of the great whales which once cruised our shores, as well the large halibut and the great schools of "whale cod" that once exerted major shaping influences on the Gulf of Maine ecosystem. And besides, none of us would volunteer to be the first to vacate our homes and livelihoods to "remove" to somewhere else for the sake of a vision of the future based on a vague notion of past plenty.

So what should we do? At a minimum, we should support and encourage those groups of fishermen who see themselves as part of a larger ecological framework in the Gulf of Maine, who agree to limit some fishing activities to help create a more abundant system for us all, themselves included. Self-interest is, after all, the most powerful human motivation; arguably, it's more worth betting on than an emotionally sterile appreciation of Nature itself. The good news in the Gulf of Maine is the existence of a small but robust conservation ethic among even the most individualistic fishermen. How else does one explain the

Russian scallop boat after one day's haul

GARY COMER (2)

Returning a petrel to the light airs of the Pacific

astounding agreement among all groundfishermen in Maine who supported a five-year ban on the taking of groundfish by any method within 2,900 square miles of state territorial waters during the three-month spawning season? How else does one explain the gradual "build down" of lobster traps toward a limit of 800 traps in state waters, supported by a very substantial majority of lobstermen? Or another handful of conservation methods, designed by lobstermen and implemented in Gulf of Maine waters? These developments are evidence of a stewardship ethic, or at least the beginnings of one, in this region's culture.

Marine protected areas can play an important role. Such areas need not be large no-fishing zones as in the Kurils and Aleutians. Here in the Gulf of Maine we need a network of smaller, easily identified and easily enforced no-take zones, designed to serve as baseline areas where scientists will, for the first time ever, be able to separate the changes in the system that result from Mother Nature from those caused by humans — from shoreside industry and human waste disposal to recreational activities or commercial fishing. The costs of creating and maintaining such a network of small areas should be borne widely throughout our society. No matter where within the vast Gulf of Maine system we may live and work, we have all played a part in altering its conditions. But without a network of marine-protected zones we are likely to continue the endless debates about what activities are causing changes in the ecosystem, and which sector should be primarily responsible for undoing the damage.

The Kurils and the Aleutians offer contrasting examples of attempts to protect marine life and can serve as models to teach us much about the Gulf of Maine.

2000

Spirit of the Grass

Katie Vaux

Zoe Lucas sits quietly in the dune grass. It rises and falls around her in waves of mottled green. It is lush and sharp, and conceals the crest of a 70-foot sand dune that rises steeply out of the Atlantic Ocean. Nearby, six wild horses graze: a black stallion, a bay-colored juvenile male, two chestnut mares and two chocolate foals. Zoe silently observes them as they migrate toward and away from each other, dictated by the direction of their grazing and their proximity to one another.

A curious mare slowly approaches Zoe. She is a light chestnut color, a young mother with a round belly, a plump rear and a thick blaze of white running down her face. She puts one hoof in front of the other, slowly but deliberately, until she is within inches of Zoe. Then gently, she lowers her neck and touches Zoe 's face with her nose. The two of them stay there for a long moment, exploring each other 's personal space.

COURTESY OF ZOE LUCAS (4)

This kind of encounter is not routine for Zoe, but it is not uncommon either. She has lived the last 30 years on Sable Island. A self-taught biologist, she abandoned a successful career as a goldsmith in Halifax to live on a sparsely inhabited sand bar 160 kilometers southeast of Nova Scotia....

These days, getting to Sable Island is no small task. Anyone traveling to Sable must get permission from the Canadian Coast Guard, and prove they have just reason to be visiting the island and adequate ground support once there. The island has no medical resources, fire department or emergency crew; an accident on Sable can quickly become an emergency....

Historically, records show that getting to Sable Island wasn't all that difficult; the challenge was getting there intentionally. Two hundred twenty-two shipwrecks have been recorded on Sable since 1801, and the island is known as "The Graveyard of the Atlantic." Historians estimate the number of shipwrecks over time at about 500, with thousands of lives lost.

The most influential factor lending Sable to shipwreck is its location directly in the confluence of two great ocean currents: the frigid Labrador Current, and the Gulf Stream, whose tropical waters warm the eastern coast of North America, Britain and Norway. This confluence creates intensely thick and damp fog, as well as strange localized currents. Strong winds blow in all directions, resulting in extremely treacherous and unpredictable conditions....

For me, as for some of the early settlers who came to the island 300 years ago, going to Sable is the adventure of a lifetime. Flying out on a fixed-wing propeller plane, it's an hour and a half before the island rises out of the open Atlantic.

Zoe Lucas has devoted her life to the study of one of the last known populations of wild horses to live without human intervention.

From nowhere, its enormous stark white dunes come out of the sea, towering over the deep blue water. As the plane begins to descend, thousands of sun-bathing seals scatter into the water. From the air, I can see bands of wild horses grazing, a few small structures on the island and the crumbling foundations of other structures lost over time to wind and sand. As we come closer and closer, I realize that I cannot see a single tree, or boulder, or handful of brown earth. Just sand and grass. When I step out of the loud, sterile body of the plane, my head is encased by the salt-soaked air and the roar of waves, glistening and pounding on the shore beside me....

Zoe greets me. She is a small woman, almost elfin in appearance, with bright eyes and a ready smile. She is to the point, and happy. Apart from walking, her only means of transportation on the island is a one-person ATV. Stray bones, bird wings, a horse skull and a knife stick awkwardly out of a crate strapped to the back.

"I 've been here 30 years," she tells me. "My entire adult life." She straddles the ATV and turns the key to start it. "I don 't even have my driver's license!"

Along the beach, the island begins to reveal itself to me. To our left, the crashing surf. To our right, massive dunes that hide the sun. They are capped by dune grass, and prickly, low-growing vegetation. Shifted and shaped by the wind, the dunes end abruptly, falling 30 feet down into sculpted white sand blow-outs, valleys created as dunes pull away from each other. These white valleys are hot and bright and difficult to walk through. Sand gives way with every step. Surrounding me are dozens of sun-bleached skulls, ribs and vertebrae, slowly being buried by sand. Scattered among them are the remains of human objects lost to the tide and found to Sable.

On the far end of this blow-out, a band of horses gather by a water hole. Using their hooves, they dig into the freshwater "lens" that lies underneath the island and provides drinking water for both horses and humans. Slowly, they meander towards the beach, passing us without pausing.

Back on the beach, we stop at the decomposed body of a horse. It died lying on its side in the sand, and has not been moved since. After a horse dies, Zoe notes its location, and leaves it for gulls and maggots to eat the flesh and digest. Now all that is left is the bones and the tough, dry leather fastening them together. Sand has filled in the depressions that used to be round with muscle. This is the collapsed shell of a stallion.

Zoe hops off the ATV with a short, sharp knife and cuts the skull away from the body. She places it in a woven plastic bag, and straps it onto the front rack of her vehicle. She will take measurements of the skull and add the data to her long-term study of the horses.

"This horse was a band stallion. I don 't know how he died. One day he was in fine health, and the next day I found him here, dead," she says. This kind of sudden, inexplicable death is very uncommon, she adds.

Zoe talks about the horses in both very personal and very scientific terms. She discusses their individual personalities while pointing out a mare she doesn 't expect to survive the winter. She makes no assumptions about their capabilities or limitations.

Zoe first came to Sable Island in the early 1970s, while studying painting and goldsmithing at the Nova Scotia College of Art and Design.

"I was stunned by how beautiful it was," she recalls. "Everything was reduced to very essential greens and blues, with splashes of color from bands of horses and colonies of seals. But I didn't want to paint it. The island is far more interesting than a painting. I didn't just want to observe it. I wanted to integrate myself into it."

After making four or five trips to the island as a volunteer, Zoe began working summers on a dune restoration project funded by Mobil. Over

time, she found herself observing the horses, and compiling the data she recorded. This led into other work, including projects on Harp and Hooded seals, shark predation and oiled bird surveys. She continued working part-time on the island while earning her masters degree in goldsmithing, and later teaching the same subject at the college. By the early 1980s, she had spent nearly ten years working part-time on the island. That was when she made the decision to leave her steady job to live year-round on the island and pursue her independent research.

"I was not thinking I would be here for good," she recalls. "There was no big plan. I just wanted to be here. I was more interested in Sable than anything else. I couldn't resist the intellectual challenge it presented: to learn directly from the source, instead of from books." She pauses, then continues. "I was attracted to its gentleness in space; not cluttered with the activities of people."

The first people to arrive on the island were the Portuguese, who brought cattle and pigs here in the 16th century.... Between 1755 and 1760, Thomas Hancock, a Boston merchant, brought a large herd of horses to the island (animals that most likely belonged to the recently expelled Acadians). When Hancock died in 1764, his plan to ship the pasturing horses to the West Indies had

Eventually, man-made structures fill with sand.

241

not come to fruition, and the horses stayed.

By domestic standards, Sable Island horses look shaggy. Their manes are long and tangled and bleached by the sun. Their winter coats are thick and soft and disheveled. They are descended from a collection of breeds, and Zoe attributes their compact, stocky build to the need for the most efficient way of maintaining body heat.

The number of horses on the island has ranged, over the years, from 200 to 400. They have neither died off nor overpopulated. Currently, there are about 400 horses, divided among 45 bands. Bands can be as small as one stallion and one mare, but they can also be much larger. Family bands can be made up of one or two stallions, a number of mares, foals and juvenile males. Bands sometimes split when a juvenile male steals a mare for himself and creates a new band. Stallions often battle for control of a band.

In the past, Sable Island horses were routinely rounded up, and those who could be captured were sent to the mainland to be sold. When Alexander Graham Bell visited the island in 1898, vainly searching for friends whose boat was wrecked on the island, he took three horses back with him. In the past few decades, islanders can recall visitors attempting to lasso horses, trying to ride them, trying to feed them, and even trying to take them home.

In the late 1950s, the government of Canada voted to sell off the horses of Sable Island. After much controversy, the decision was reversed, and the government instead passed laws protecting the horses from all human interference....

We continue on along the north beach, moving farther and farther east. The dunes here rise higher and fall more steeply than others on the island, reaching 90 feet before diminishing into a spit of sand that trails on for five kilometers before disappearing into the sea.

Nearing the eastern tip of the island, we have left the dunes behind and cruise out over the spit, which lies only a few meters above the water, and is less than half a kilometer wide. I can see both "sides" of the island in one glance, as well as parts of the spit where the tide has washed over, carrying away any marine litter or animals in its way. Once while cruising along this spit, a high tongue of surf washed under Zoe, floating her ATV and taking it out to sea. She escaped the vehicle before she too was lost to the tide.

Stretching nearly five kilometers, this part of the island has a startling psychological impact on me. I feel vulnerable, desolate. There is no grass out here, and no horses, just the howling wind that steals warmth from my body. The beach is streaked with veins of red mineral sands that from a distance look like thick rivers of blood dripping across the white spit and into the water. It is a clear day, but the enormous, intimidating dunes have suddenly disappeared in the distance until all I can see is this spit of sand, and the thousand ways a person like me could die out here.... I cling tightly to Zoe, and her simple presence anchors my psychological composure.

We stop often along the spit, as Zoe scans the beach for oiled seabirds, part of a project funded by Mobil. We also stop each time we pass a deflated child's balloon, which she picks up and tucks into her crate. This is part of an independent marine litter survey she has been conducting over the past 3 years. "I want kids to know what happens to their balloons. They don't have many opportunities to make a difference in their world, but this way, they can at least make the decision whether they want to have balloons at their parties or not," she says.

Being the only land for hundreds (if not thousands) of miles in almost every direction, Sable catches enormous tides of human refuse. Marine litter on Sable consists of everything from pop cans and garbage bags to television sets, computers, stoves, freezers, channel markers, marine flares, boat equipment and the occasional message in a bottle. Zoe once found a prosthetic leg buried vertically in beach sand, the bottom of its foot facing the sky.

"I was less afraid of the prosthetic leg than I feared somebody might be attached to it — which they were not," she says. "I kept that leg for a while. I would leave it somewhere in a room, and somehow it would move to another spot, although I was always sure I hadn't moved it. I finally got rid of it."

In times past, more than refrigerators and buoys washed ashore on Sable. It was people, and the vessels that carried them...In 1801, the Governor of Nova Scotia, alarmed by reports that "a man and woman of wicked character" were inhabiting Sable "for the infamous, inhumane purpose of plundering, robbing and causing shipwrecks," sent a scout to investigate the island. Instead of finding depraved marauders, the scout sent back word that Sable was inhabited by a few poor souls, and that the island was in desperate need of people willing to aid those stranded by shipwreck. That year, the Humane Establishment was founded, and would spend the next 157 years saving the lives and cargoes of 222 shipwrecks on Sable.

The lifesavers and their families were expected to feed themselves on Sable, and grew vegetables and raised cattle, pigs and poultry. They grew their own hay, and gathered driftwood and ship timbers for fuel. Three lifesaving stations were spread across the island, and a boarding school was estab-

lished at Station 2 for the island children to attend, if and when a tutor was available... In 1958 the Humane Establishment closed its doors. When the last families to live on the island left in the late 1960s, the human presence on the island was focused purely on conservation efforts....

The dune restoration project that first brought Zoe Lucas to Sable in 1971 evolved into a lifetime of living on an extraordinary island, among a unique population of horses....These days, she is focused on a long-term genetic study of the horses, collecting the skulls, teeth and bones for measurements and analysis. That is what has brought us to a marshy area near one of the island 's freshwater ponds. Walking towards the pond, we suddenly come upon an extraordinarily verdant patch of grass and roses. In the middle of it is a horse, almost completely decomposed.

Almost.

Next thing I know, I am on my knees, digging with my hands through thousands of empty black maggot cases, and the remains of the horse 's stomach — including what was in it when he died. Among the bones and flies there are blond curls of his mane, still long and beautiful and tangled.

Zoe and I work together, pulling out each and every bone with great care. When we are finished there is barely anything left. Within a few weeks it will be entirely grown over, leaving no evidence of the horse that lived and died here. So this is what happens when a horse dies, I thought.

But what happens to the spirit of the horse? I look over my shoulder to see a young foal with a mouthful of green blades, and I know that I believe Zoe when she tells me: It becomes the spirit of the grass.

As one of an increasingly small number of such places, Sable Island is still a place where the natural rhythms of life and death remain relatively unchallenged by human intervention. Zoe Lucas has spent a lifetime in quiet observation, and contemplation, of these rhythms. The horses here, and the seals; the birds and the sharks; the endless stretches of grass and surf; this is what Henry Beston meant when he wrote about "other nations."

For Zoe, living on Sable Island has meant living as a nation of one. And to live respectfully with nations she cannot make assumptions about, Zoe has instead lived among them as though she herself were merely a ghost who prowled the dunes.

2002

GARY COMER (3)

Summer Ice

Philip Conkling

"Greenland," wrote artist-adventurer Rockwell Kent in 1931, "that small part of which is not buried under the eternal ice cap ... is a stark, bare, treeless land with naked rock predominating everywhere." While Kent's description reflects Greenland's bleakness, it doesn't do justice to the beauty one also encounters on and around this vast island. Its mountain glaciers and floating ice, moving as they melt in the high Arctic sun, are spectacular examples of powerful, dangerous natural forces at work. Ten percent of the icebergs in the entire North Atlantic originate in a single bay in West Greenland.

In July 2000 and 2001, I joined the expedition vessel TURMOIL as it explored the northern reaches of the North Atlantic beyond the Arctic Circle. The first voyage crossed Greenland Sea from Iceland and ultimately reached the north coast of Spitsbergen. The next year TURMOIL ventured up Greenland's western shore. Both years we pressed north until impassable pack ice forced TURMOIL to turn back. On the 2000 the 76th parallel at Melville Bay after a thousand-mile voyage up Greenland's west coast. Along the way the vessel's owner, Gary Comer, an Island Institute member and the founder of Land's End, took the color photographs on these pages, and I kept the ship's log....

CROSSING THE GREENLAND SEA
JULY 13–15, 2000

As our vessel cruises north from Iceland, it is hard to keep track of time when there is no difference between day and night. Most of us sleep throughout the nighttime hours with the portholes covered to block out the white light while our highly competent crew moves TURMOIL, a 151-foot cruising vessel, through the Greenland Sea 'round the clock.

Throughout the day, the ship's routine goes on smoothly and unceasingly. TURMOIL's four passengers find little corners in the library, the main saloon or the bridge to catch up on reading or to send e-mails back home. The satellite coverage for regular telecommunications gets patchy north of 70 degrees latitude, so today may be our last contact with our offices and families back in Maine.

Tonight the sun will not dip below the horizon at all, but will roll around the rim like a giant yellow marble balanced on the lip of an enormous saucer until it slowly begins to ascend on the other side of due north.

POLAR ICE AT THE 80TH PARALLEL
JULY 19, 2000

After a quiet night in a fjord at the northern end of Spitsbergen we awake to gradually clearing skies. TURMOIL heads north from her anchorage with the intention of exploring the edge of the sea ice where we might see polar bears hunting for seals. As the clouds continue to lift in the shimmering morning sunlight, we begin to see mirages of ice off to starboard. A fata morgana, *a mirage that* appears in polar regions caused by air temperature and density differences, now fills the entire eastern horizon, looming up like a vertical wall through the binoculars. This can only be the edge of the pack ice, although it looks like the edge of an ice escarpment — a barrier no one would ever be able to cross.

Soon we are nosing through what the captain calls a field of "bergy bits," little pieces of ice dotting the surface. A mile farther ahead we swivel into the edge of pack ice. When we arrive, Gary asks the crew to launch one of the ship's small boats so we can maneuver through the floating bergs and get a closer look at the ice field swaying on gentle Arctic swells. Meanwhile puffins careen by overhead, seemingly flying through the rigging of TURMOIL. The sun glints silver off the shifting ice pack. The sound of the pack ice is like a rushing wind off in the distance, as the heave and surge of the sea lifts and lowers a thousand square miles of restless ice that shifts uneasily against itself like a huge, moaning diaphone. It is the dull roar from the lair of the ice king.

A short while later, back aboard TURMOIL in the pilothouse, there is a brief celebration at the GPS station as we cross the 80th parallel. Only 600 miles to the Pole! For one mad moment, the same thought flashes across all our minds; maybe we will get lucky and find a long enough lead through the ice pack to make a dash for the Pole and get to the top of the world! Only for a moment, though. Who in his right mind would risk this boat for such an ice-crazed dream? But still...

MELVILLE BAY
JULY 16, 2001

The early morning light collects in pure pools of silver across the eastern edge of Melville Bay. Not a cat's paw of wind disturbs the surface of the sea in a breathless inhalation of dawn. The vaporous eastern sky is the palest iron blue while the horizon's clouds to the west merge imperceptibly with the shimmering edge of the water. No distinct line separates sea, air, ice or light where the curvature of the heaven meets the arc of the circumpolar sea. It's as if the elements shift from one semi-solid state to another. Nothing here is fixed; nothing seems certain.

We have gone as far as we can go for the time being on our northerly course and turn west paralleling the broad curve of the land. To add to the mystery and eeriness of the place, long, luminous green tendrils of some kind of life form begin appearing in the water a meter or so deep, looking pale and shining at the same time. These can only be streaks of plankton, but this thought raises as many new questions as answers. Why they are arranged in long lines; what pale fire lights these numinous streaks?

There is room for TURMOIL to maneuver among the floes, so onward through the ice we press and pry. At one point, our bow wave brushes up against a piece of ice the size of a backyard swimming pool. As the wave disturbs the ice, it breaks into a few pieces and then it dematerializes. It simply becomes water before our wondering eyes. Now that we have arrived at the pack ice north of the 75th parallel, life has suddenly exploded around us. Flocks of murres and dovekies fly by as if they were pellets shot from a shotgun. Fulmars tilt past, dipping one wing and then the other to navigate the sinuous ice margins. And then we begin to see seals. Gary spots what they believe to be an old bull hooded seal from the shore boat. Now from the bridge of TURMOIL we see four harp seals sprawled on a large flow.

The further we twist and turn our way into the ice flows, the more abundant becomes the marine life. The base of the food chain here is a brown algae that attaches to the bottom of the ice, photosynthesizing through the translucent light. Various zooplankton graze on this food supply and become food for larger shrimp like creatures which feed the fish and seabirds which feed the seals and whales. The idea that the ice is lifeless is a temperate zone concept that is turned on its head here in the Arctic.

As we proceed on our meandering northwesterly course, a dilemma stares us in the face. If we assume that the satellite ice map of yesterday showing the extent of the pack ice 100 miles across our course has somehow greatly dissipated, we might be able to traverse Melville Bay at 5-6 knots to reach Cape York in another day of strain on the crew and vessel. Another 75 miles beyond Cape York is Thule, our ultimate goal. We steam on winding through the magnificent narrow channels flaked by icebergs and shoals of pancake and brash ice at half speed, not quite willing to come this close to the legendary destination of Thule, and just let it go. In the still glassy diapauses of the afternoon, we relaunch the shore boat to be nearer to the ice and water and sea life. The sun burns hazy holes through the high cloud cover. The wind inhales deeply and holds its breath. Every sensation is full and rounded. If only this would never end.

Late in the afternoon a final set of satellite ice maps seals the decision we had all silently reached earlier. TURMOIL nudges up to a floating ice pan and shudders to a halt. Slowly, painfully she backs down and thrusts her bow around in a broad arc. We rejoin the vessel and the crew, load the shore boat back on the deck and facing aft, begin retracing our course to the south.

LESSONS FOR THE GULF OF MAINE

Several weeks after returning from TURMOIL's July 2000 voyage, we were stunned, like many others, to read a front-page story in the *New York Times* reporting that a Russian icebreaker, YAMAL, had left Spitsbergen with a group of tourists in mid-August headed for the Pole. When they got there a week later, instead of ice, they discovered nothing but open water. There were several American scientists leading the tourist expedition aboard the YAMAL, including Dr. James McCarthy, an oceanographer and director of the Museum of Comparative Zoology at Harvard University.

McCarthy said that he had been at the Pole six years earlier and had encountered a completely frozen sheet of ice. This time, he reported, he had never seen so much open water in the polar region. The Captain of the YAMAL said he had been making this passage for the past decade and generally had to cut through an ice sheet six to nine feet thick. Other reports from prominent polar scientists published a few weeks later pointed out that approximately ten percent of the polar Arctic region is open water, since the ice pack constantly shifts in winds and currents to create pockets of open ocean, and that open water will appear from

time to time at the Pole.

Despite such disagreement, the evidence is mounting that an unusual amount of melting is occurring in the polar regions of the North Atlantic. What effects this might have on the rest of the ocean, including our small window on the world at the edge of the Gulf of Maine, is anyone's guess.

Closer to home, a story from last summer, perhaps insignificant in the global scheme of things, is a reminder of how interconnected life in the ocean can be. As part of the Penobscot Bay Marine Collaborative, the Island Institute administers an interdisciplinary team of scientists, fishermen and managers trying to understand the dynamics of lobsters in the bay. Part of the project is to place college graduate "Island Fellows" and interns

TURMOIL, a 151-foot exploration vessel, ranges the world gathering scientific knowledge about oceans and polar regions.

aboard lobsterboats to collect information from cooperating fishermen. Lobstermen know that every year the habits of the world's most favored crustacean will be somewhat different than the previous year. Since anyone can remember, lobstermen in the spring have begun to set traps close to shore to intercept lobsters as they crawl into shallower, warmer waters to shed their old shells and breed. But 1999 was dramatically different. Fishermen set their gear as always, but catches in April, May and early June were not just low, especially in the Midcoast — they were almost nonexistent. Already nervous because of the collapse of lobster population in Long Island Sound, N.Y., Maine fishermen feared the worst.

Then, beginning in early June, a strange thing happened. Lobstermen began catching shedders, very large numbers of them, between three and four weeks earlier than normal. And many of the lobsters with new shells were caught in deeper water than usual. The evidence is anecdotal and therefore "unscientific," but a picture emerges: The early part of the lobster season was dramatically different from what many established fishermen had ever observed.

What we know is that the earth's surface temperature, including the temperature of the sea's surface, has increased by about one degree since the late 19th century, and that the 1990s have been the warmest decade on record. The difference in average global temperatures between a full-fledged ice age and our present interglacial climate is only about five degrees, so this global temperature increase is cause for concern, if not alarm. We also know from detailed recent measurements that the southern half of the Greenland ice sheet — the

second largest ice sheet in the world after Antarctica — has shrunk substantially in the last five years. Greenland is losing ice at a rate comparable to the size of Maryland covered by a foot of ice melting per year. Finally, measurements by the U.S. Navy, from submarine logs of voyages under the Pole, also reveal that the ice layer is substantially thinner than it used to be.

No one has definitely linked the melting of ice in the high Arctic to the production of greenhouse gases released by the burning of fossil fuels, but it is worth asking what kind of worldwide experiment is now ineluctably underway. Global warming scientists used to model a doubling of greenhouse gases in the earth's atmosphere and try to interpret the results. Newer projections make a tripling of the volume of greenhouse gases ever more likely during the 21st century.

Those with their hands on the levers of government, at least on this side of the North Atlantic, have taken a "wait and see" approach to the debate over whether global warming is real and a cause for concern. Because scientists are unable to predict with any certainty the timing and magnitude of global climate change, nor even whether it will produce new deserts or new ice ages, some might think this strategy has the prudence of Solomon. But if you view the doubling or tripling of heat-trapping gases in the atmosphere to a global game of Russian roulette, especially when the outcome is so unpredictable, the strategy looks less like Solomon's and more like Nero's.

2001; 2002

GARY COMER (2)

Peter Ralston and Philip Conkling, founders of the Island Institute

PAIGE PARKER

"A PUBLICATION CAN SHAPE AN ENTIRE ORGANIZATION," we wrote in the introduction to the 2003 *Island Journal*, marking 20 years of publication. This magazine was to be "a meeting place for ideas, a forum for discussions, a venue for poetry and literature, a showcase for photography and the visual arts."

Looking back, we believe that *Island Journal* has achieved what it set out to do. But it has done so largely in the context of a larger enterprise, the Island Institute, which dedicated itself to community development that was, like *Island Journal*, island-centered. Where *Island Journal* told the stories of islanders, the Island Institute did what it could to ensure that more would remain of the islands of the Gulf of Maine than ruins, recollections and the written record. The Institute's mission is to foster community sustainability in many forms: in fisheries, in schools, in local government, in economic and social life.

In pursuit of sustainability the Island Institute has helped island communities defend themselves against inappropriate development and other incursions from the mainland. It has worked with island schools and their teachers to ensure that island children receive the education they deserve. It has worked with island landowners to protect forests from fire and disease; it has promoted economic development at a scale appropriate to small communities; and it has been a bridge between fishermen and scientists. Most important, the Island Institute has been a forum, a facilitator, in the often-contentious dealings between island communities and the mainland authorities — towns or counties that include islands but often forget about them — and the state and federal governments whose prescriptions for schools, environmental protection, taxation and fisheries may not fit island settings.

Island Journal should be seen in this larger context. Like its sister publication, *Working Waterfront*, it's the storytelling part of the Island Institute, illuminating island life through the eyes of those who live it, enabling islanders to share their stories with the wider world.

The Editors

James Acheson is a professor of anthropology at the University of Maine. His books include *The Lobster Gangs of Maine.*

Milton Achorn was born in Charlottetown, Prince Edward Island, in 1923 and died there in 1986. He was known in Canada as "The People's Poet."

Jan Adkins is author of *The Craft of Sail, Wooden Ship* and other books.

Ted Ames is a commercial fisherman and writer in Stonington, Maine.

Edgar Allen Beem is an art critic and author of *Maine Art Now.*

Peter Benchley is a freelance photographer on Nantucket.

Elizabeth Bishop was awarded the Fellowship of The Academy of American Poets in 1964.

Philip Booth has published his poetry in *The New Yorker*, *The Atlantic Monthly*, *The American Poetry Review*, *Poetry* and *Denver Quarterly*.

Cynthia Bourgeault was editor of *Island Journal* from 1987 to 1993.

Mike Brown is a freelance writer whose columns appear in several Maine newspapers.

Tom Cabot was a founding member of the Island Institute and an early advocate for the preservation of Maine islands.

Steve Cartwright is a freelance writer who contributes regularly to *Working Waterfront, Island Journal* and other publications.

Candace Cochrane is Director of Education at the Quebec-Labrador Foundation.

Gary Comer is the founder of Lands'End and a founding member off the Island Institute.

Philip Conkling is President of the Island Institute.

David Conover is a writer and filmmaker in Camden, Maine.

Christopher Crosman is director of the Farnsworth Art Museum in Rockland, Maine.

Al Diamon is a freelance writer in Portland, Maine.

Sandra Dinsmore writes regularly for *Working Waterfront, Island Journal* and other publications.

Bill Drury taught at the College of the Atlantic in Bar Harbor, Maine.

Jeff Dworsky is a photographer and fisherman who lives in Stonington, Maine.

Keith Eaton was an Island Institute fellow on North Haven.

James Essex was an Island Institute Fellow on Peaks Island.

Mike Felton is a former Island Institute Fellow on Vinalhaven who now directs the Institute's education programs.

Hortense Flexner worked as a writer for Curtis Publishing from 1923 to 1929, while she established herself as an active and published poet. She also taught at Bryn Mawr and Sarah Lawrence.

John Fowles is a novelist whose books include *Islands, The Magus* and *The French Lieutenant's Woman.*

Maggie Fyffe is secretary of the Isle of Eigg Heritage Trust.

Bridget Besaw Gorman is a freelance photographer based in Portland.

Emily Graham is a school librarian and former Island Institute Fellow on North Haven.

Nancy Griffin writes for numerous Maine publications including *Working Waterfront* and *Island Journal.*

Karen Roberts Jackson of Green's Island writes regularly for *Working Waterfront* and *Island Journal.*

Hope, Dylan and **Tristan Jackson,** the children of Karen and Mark Jackson of Green's Island, described their experiences in Nunuvut for *Outside* magazine and *Island Journal.*

Dana Leath is a Senior Fellow at the Island Institute, based in Portland. She was an Island Institute Fellow on Long Island.

Dean Lunt is a publisher, a former newspaperman and author of *Hauling by Hand.*

Robert Lowell is considered by many to be the most important poet in English of the second half of the twentieth century.

Cabot Martin co-founded Vinland Petroleum Inc., the first Newfoundland and Labrador-based oil and gas exploration company. Later he founded Deer Lake Oil and Gas Inc.

Nate Michaud is a former Island Institute Fellow who now manages the Institute's programs.

Ben Neal is the former Marine Programs Manager for the Island Institute.

George Oppen received the Pulitzer Prize for *Of Being Numerous* (1968).

John Patriquin is a photographer for the Portland Press Herald.

Amy Payson lives and writes in South Thomaston, Maine.

Chellie Pingree represented North Haven in the Maine Legislature, ran for the U.S. Senate and is today national director of Common Cause.

David Platt is director of publications at the Island Institute.

George Putz was founding editor of *Island Journal.*

David Quammen is author of *Song of the Dodo* and writes for *Outside* magazine.

Peter Ralston is Executive Vice President of the Island Institute.

Kathleen Reardon was an Island Institute Fellow on Islesboro.

James Rockefeller is a longtime seasonal resident of Vinalhaven.

Jeanne Rollins lived and wrote on Monhegan for many years.

May Sarton's collections of poetry include *Coming Into Eighty, Collected Poems: 1930-1993, Halfway to Silence, A Private Mythology, The Lion and the Rose* and *Encounter in April.*

Susan Hand Shetterly is a Maine-based essayist who contributes frequently to *Island Journal.*

Harry Thurston was awarded a Canada Council Explorations Grant in 1985 to write *Tidal Life: A Natural History of the Bay of Fundy.* Other books include *The Sea Among the Rocks* and *The Nature of Shorebirds.*

Joe Upton is author of *Alaska Blues, Journeys Through the Inside Passage* and *Runaways on the Inside Passage.* He is a summer resident of Vinalhaven.

Katie Vaux is a freelance writer in Nova Scotia.

David Weale is a professor of Canadian and Prince Edward Island history at the University of Prince Edward Island.

Colin Woodard is author of *Ocean's End: Travels Through Endangered Seas,* and a columnist for *Working Waterfront.*

Page iii: "Matinicus 68°55′W – 43°52′N" by Philip Booth, from *Islanders*, Viking Press, 1986;
The Fish Wharf, Matinicus Island, George Bellows, 1916. Portland Museum of Art, bequest of Elizabeth B. Noyce.

Page viii: *Reefer*, Andrew Wyeth, 1977: courtesy of the artist.

Page 11: *Breakfast at Sea*, Jamie Wyeth, 1984, courtesy of the artist.

Page 5: *"The Island,"* by Milton Achorn, reprinted with the permission of the estate of Milton Achorn.

Page 13: *"The Woods After Rain,"* by Hortense Flexner, from Poems for Sutton Island, High Loft Press, 1983.

Page 18: "Siren Call," by John Fowles, from *Islands*, Little Brown, 1978;
Bronze Age, Jamie Wyeth, 1967, courtesy of the artist.

Page 21: *The Wind*, Jamie Wyeth, 1999, courtesy of the artist.

Page 27: *The Islander* (detail), Jamie Wyeth, 1975, courtesy of the artist.

Page 38: "Islanders," by Philip Booth, from *Matinicus*, Viking Press, 1986.

Page 52: *The Coot Hunter*, Andrew Wyeth, 1941, courtesy of the artist. Art Institute of Chicago,
Olivia Shaler Swan Memorial Fund, 1943.

Page 53: "A Farewell," from *Letters for Maine*, copyright 1984 by May Sarton, W.W. Norton & Co., Inc.

Page 55: *The Odom Brothers*, Jamie Wyeth, 2001, courtesy of the artist.

Page 56: *The Morris House*, Port Clyde (detail), N.C. Wyeth, 1935,
collection of the Farnsworth Art Museum, Bequest of Mrs. Elizabeth B. Noyce, 1997.

Page 63: "Penobscot," by George Oppen, from *Maine Lines*, 1996, compiled by Richard Aldridge, Harper-Collins Publishers Inc.

Page 73: *Crew Neck*, Andrew Wyeth, 1992, courtesy of the artist.

Page 83: From *The Country of the Pointed Firs*, Sarah Orne Jewett, Houghton Mifflin, 1910.

Page 88: *Dark Harbor Fishermen* (detail), N.C. Wyeth, 1943, Portland Museum of Art.

Page 93: "It Was Good and It Was Enough," by Joe Upton, from *Amaretto*, International Marine, 1986.

Page 98: "Dragging," by Philip Booth, from *Relations, Selected Poems 1950-1985*, Viking Penguin, Viking Books USA.

Page 138: *Teels Island*, Andrew Wyeth, 1954, courtesy of the artist. Image provided by the Art Institute of Chicago.

Page 142: *Iris at Sea*, Jamie Wyeth, 1994, courtesy of the artist.

Page 145: "Water," by Robert Lowell, from *For the Union Dead*, Farrar Straus & Giroux, copyright 1992.

Page 150: *Beach Flowers #2*, Fairfield Porter, 1972, Portland Museum of Art.

Page 151: *Spring Islands* (detail) Eric Hopkins, 1988, courtesy of the artist.

Page 152: *Airborne*, Andrew Wyeth, 1996, courtesy of the artist.

Page 153: *The Herring Net*, Winslow Homer, 1885, Art Institute of Chicago.

Page 154: *Maine Coast*, Rockwell Kent, 1907, collection of the Farnsworth Art Museum,
bequest of Mrs. Elizabeth B. Noyce, 1997.

Page 155: *Orca in Winter,* Jamie Wyeth, 1990, courtesy of the artist.

Page 157: *Cold Ruffles II,* Brita Holmquist, courtesy of the artist;
Decade Autoportrait, Robert Indiana, 1982, courtesy of the artist.

Page 158: *Monhegan Headland* (detail), Reuben Tam, 1968, courtesy of the artist.

Page 159: *The Dock,* Fairfield Porter, 1974-75, courtesy of the Portland Museum of Art.

Page 160: *Late Afternoon,* Rockwell Kent, 1906, courtesy of Jamie Wyeth

Page 161: *Winter, Monhegan Island,* Rockwell Kent, 1907, The Metropolitan Museum of Art,
George A. Hearn Fund, 1917.

Page 162: *Toilers of the Sea,* Rockwell Kent, 1907, New Britain Museum of American Art,
Charles F. Smith Fund, 1944.

Page 164: *Light Station* (detail), Jamie Wyeth, 1992, courtesy of the artist.

Page 166: *Meteor Shower,* Jamie Wyeth, 1993, courtesy of the artist.

Page 167: *Dr. Syn,* Andrew Wyeth, 1981, courtesy of the artist.

Page 169: *Southern Light,* Jamie Wyeth, 1994, courtesy of the artist.

Page 171: *Bees at Sea — Study #1,* Jamie Wyeth, 1994, courtesy of the artist.

Page 178: "North Haven: In Memoriam, Robert Lowell," from *The Complete Poems,* 1927-1979
by Elizabeth Bishop, copyright 1979, 1983 by Alice Helen Methfessel, Farrar, Strauss & Giroux LLC.

Page 181: *Urchin* (detail), Jamie Wyeth, 1999, courtesy of the artist.

Page 182: *Ravens in Winter,* Jamie Wyeth, 1996, courtesy of the artist.

Page 184: *Offshore Raven* (detail), Jamie Wyeth, 1996, courtesy of the artist.

Page 185: *Saltwater Ice,* Jamie Wyeth, 1997, courtesy of the artist.

Page 186: *Ravens, Standing,* Jamie Wyeth, 1996, courtesy of the artist.

Page 187: *Untitled Study,* Jamie Wyeth, 1996, courtesy of the artist.

Page 189: *The Rookery,* Jamie Wyeth, 1997, courtesy of the artist.

Page 220: "I, Milton Achorn," reprinted with the permission of the estate of Milton Achorn.

ACKNOWLEDGMENTS

ISLAND JOURNAL IS A COMPLEX PROJECT. This book and the 20 issues of the
magazine that preceded it were made possible by the hard work of dozens of people,
as well as the generosity of foundations and individuals, some of whom prefer
to remain anonymous. The Engelhard Foundation, which has supported this unique
publication for many years, deserves special mention.

People whose efforts have been particularly significant in *Island Journal*'s
history include George Putz and Cynthia Bourgeault, its first editors; Anne Leslie
and Esme McTighe, copy editors; Penmor Lithographers and The J. S. McCarthy Co.,
its printers for many years; Michael Mahan Graphics, longtime designers of *Island Journal*;
Sandra Smith and Michael Herbert, of the Island Institute's publications staff, who
have been responsible for distributing *Island Journal* to dozens of stores where it is
sold; and the Membership Department at the Institute, which makes certain the magazine
reaches more than 4,000 members annually. In addition, many people and institutions,
including the Farnsworth Art Museum, its Wyeth Center and the Brandywine River Museum,
have loaned artwork and historic images; we appreciate their trust and generosity.

Special thanks to Allison Philbrook, who re-typed many pre-digital stories for
this book, and to Melissa Hayes, who proofread the whole thing.

Finally, we salute all the freelance writers, artists and photographers whose
ideas and contributions are the stuff of which this magazine is, literally, made. Without
their work, compensated at embarrassingly low rates for many years, *Island Journal*
would never have achieved what it has.